NUCLEAR ENERGY AND ETHICS

NUCLEAR ENERGY AND ETHICS

edited by
Kristin Shrader-Frechette
University of South Florida

WCC Publications, Geneva

Cover design: Rob Lucas

ISBN 2-8254-1035-7

150, route de Ferney, 1211 Geneva 2, Switzerland

Design and Layout by "Spot-On" Publishing Services

Printed in Switzerland

CONTENTS

ACKNOWLEDGMENTS

In October 1989, members of the Kimbanguist Church opened their hearts and their facilities in Kinshasa, Zaire to an internationally distinguished group. Welcoming scholars, government policymakers, and representatives of many religions from all over the globe, the Church hosted the World Council of Churches' Consultation on Nuclear Energy. Without the generosity and hard work of Dr. Bena-Silu and the other members of the Kimbanguist Church, the consultation would not have been possible. Because of the example and the religious commitment of our hosts, participants in the consultation learned as much from our African brothers and sisters, the Kimbanguists, as from the experts who spoke at the meeting. We are deeply grateful to Dr. Bena-Silu and the Kimbanguist Church.

WCC staff from the Sub-Unit on Church and Society shepherded this volume, an outgrowth of the papers prepared for the Kinshasa meeting, to completion. They worked out numerous details, helped obtain copies of the manuscripts, and provided financial assistance for preparation of the essays and reports. Wesley Granberg-Michaelson, Peony Wong and, especially, Freda Rajotte, provided invaluable assistance, hard work, and good cheer.

Thanks are also due to Nancy Keen, of the University of South Florida, for her word-processing work. I owe a special debt to Daniel C. Wigley, of the University of South Florida, for acting as research assistant and word processor for the manuscript. The Philosophy Department of the University of South Florida provided funds for xeroxing, mailing, phone, and fax. One could not ask for a better team with whom to work.

Kristin Shrader-Frechette
Tampa, Florida, USA
October 1990

PREFACE

For the past two decades, the Church and Society Sub-Unit of the World Council of Churches has encouraged discussion of the various ethical issues raised by the use of nuclear energy. These issues received extensive consideration during the conference, "Faith, Science and the Future," in 1979 at the Massachusetts Institute of Technology (MIT).

Since that time, however, there have been considerable shifts in the public perception of the risks involved in nuclear power production and waste management as well as concern over the possibilities of responding to increasing global demands for energy. The Church and Society Working Committee called a consultation in Kinshasa, Zaire, for October 16-23, 1989. The purpose was to consider these developments and also to focus on theological and ethical concerns related to the production of nuclear energy. In particular, the Working Committee requested that the following issues be considered:

- questions related to the storage and handling of nuclear waste; nuclear proliferation;
- questions of safety related both to present reactors and a possible new generation of reactors; and
- the ethics of nuclear energy.

In addition, the Working Committee directed that discussion should take place within the broader framework of present and projected world energy needs in both industrialized and developing countries. It also requested that nuclear energy be evaluated as one possibility among varied energy options, each with its own environmental, social, and economic costs and hazards. Finally, the committee asked that the consultation consider the possibility of, and limits to, improving energy generating and consumption efficiency.

Most importantly, the aim of this consultation was to give attention to the role of churches as they encounter these issues. The goal was to identify important current questions deserving attention, from both a scientific and ethical point of view, rather than to propose any final position for WCC policy.

Wesley Granberg-Michaelson
Geneva, Switzerland
October 1990

Introduction: Moral Consensus and Nuclear Power

Kristin Shrader-Frechette
University of South Florida,Tampa, Florida, USA

Religion ought to make a difference in the world. Growing up in Kentucky in the late fifties and early sixties, I learned that it did. I saw that blacks and whites were separated in buses, trains, and neighborhoods, but not in our rural church. They were segregated at public schools, drinking fountains, and restaurants, but not in our parish school.

This volume argues that religion and ethics ought to make a difference in our private lives and our public policies. Often it does not. Worse still, often religion makes the wrong kind of difference in our lives, a difference created by authoritarianism, self-righteousness, or triumphalism.

Each of the essays included in this collection, commissioned by the World Council of Churches, explicitly affirms that ethics ought to make a difference in the way we view the issue of commercial nuclear fission. Ethics ought to make a difference, in part, because ethics is unavoidable. All our actions presuppose an ethical point of view. Because of this presupposition, we ought to try to uncover all our implicit ethical judgments and subject them to analysis and discussion. If we fail to realize that our actions presuppose an ethical point of view, then we do not avoid ethics. We merely make ourselves the victims of bad ethics. We may also make ourselves cowards. Silence makes us cowards, as Abraham Lincoln put it. Silence also implies consent to the status quo, including consent to the many ways that we use, and are used by, existing technologies.

If, as theologian Peter Sedgwick affirms in his chapter, we have a vocation to create a moral community of value, then we cannot remain silent about the use of nuclear fission to generate electricity. (An equally important question concerns the military use of atomic energy, but that is not the main concern of this volume.) If we wish to help create a moral community of value on the issue of allegedly peaceful or commercial uses of nuclear power, then we must engage in moral reasoning, even though there is no complete moral consensus. Our aim in presenting these chapters is to use their moral reasoning to deepen our ethics, to negotiate our differences, and hence to forge an ultimate and just agreement regarding commercial nuclear fission.

1. A Survey of the Volume

Part of the moral dissensus evident in this collection of essays arises from the fact that the various chapters have been written, respectively, by representatives of churches, governments, developing and developed nations, non-governmental organizations (NGOs), the nuclear industry, and universities. The polarization in this volume between nuclear-industry spokespersons and representatives of environmental NGOs, for example, appears almost as great as that found in society in general. After briefly summarizing the contents of each of the chapters, we shall outline the main points of disagreement.

Divided into three main sections, the volume begins with an historical survey of both commercial nuclear technology and the responses of the WCC to these developments. Section two discusses several key policy issues related to the use of commercial nuclear fission, whereas the last section deals with ethics. The book makes a clear progression from history, to policy issues, to ethics.

In the first chapter, "An Overview of the Issues Confronting Nuclear Power and the Concern of Churches," Ulrich Ratsch surveys the commercial nuclear debate within the WCC for the last 20 years. German physicist and church scholar, Ratsch points out that, as early as 1974, some groups within the WCC urged use of renewable energy sources so as to avoid the ethical problems associated with nuclear technology. Although he explains that there has never been a complete consensus within the WCC regarding commercial nuclear fission, Ratsch reveals that in 1979 the WCC passed a resolution calling for a five-year moratorium on construction of nuclear plants. At the end of his essay, he summarizes the key ethical issues that have divided nuclear proponents and opponents. The questions include the equitable distribution of energy, both globally and nationally; the relationship between energy production and economic growth, especially in developing nations; the environmental problems related to use of various energy sources; and the harms and benefits associated with commercial nuclear fission. Ratsch closes his historical survey with the observation that "hardly anyone today believes in a nuclear future."

Gordon Linsley, physicist and researcher at the International Atomic Energy Agency, presents a more optimistic history of nuclear energy and waste management than does Ratsch. In the second chapter, "Radioactive Wastes and Their Disposal," he surveys methods used to store nuclear wastes and provides a state-of-the-art account of current plans for future radioactive waste management. He also distinguishes the various types of radwaste according to degree of hazard, concentration, and lifetime. Emphasizing that waste management involves either containment of dangerous nuclides or dilution and dispersal of them in the biosphere, Linsley summarizes the functional requirements of an adequate waste-management system. He also explains that the safety of stored radioactive wastes is guaranteed by a strategy of "defense in depth," multiple barriers that together guarantee no breach of containment. Linsley admits, however, that uncertainties surround issues of the long-term safety of the wastes, the problem of future human intrusion, and transboundary considerations.

In the third chapter, Zairean physicist and church representative, Bena-Silu, likewise presents a quite positive account of the assets of commercial nuclear fission. Discussing the history, development, and operation of nuclear-fusion, nuclear-fission, and fast-breeder reactors, Bena-Silu gives a state-of-the-art survey of each of these technologies. He also sketches some of the ethical issues associated with nuclear waste, with human error, and with proliferation. His perspective on proliferation is especially significant because, as a citizen of a developing nation, he argues that the Nuclear Non-Proliferation Treaty is "very unfair," and that nuclear powers merely wish to use it to protect their hegemony. He emphasizes that developing nations themselves must make their own decisions about commercial and military uses of nuclear technology.

The fourth chapter also addresses energy strategies for developing nations. Beginning the discussion of key policy issues in section two of the book, Achilles del Callar is a university professor and a top (Aquino) government official in the Philippines. His views differ sharply with those of nuclear proponent Bena-Silu. In a detailed analysis, "The Impact and Safety of Commercial Nuclear Energy: Perspectives from the Philippines," del Callar shows how construction of one nuclear plant in the Philippines has crippled his country economically. Because commercial nuclear power is so capital intensive, the Philippine government must pay out $355,000 daily, just on the loan for the power plant. The nuclear plant loan is the largest single debt of the Philippines, which annually spends half of its budget on external debt payments. Del Callar argues that developing nations like the Philippines cannot afford such a costly technology, given their precarious financial situation and the fact that 70 percent of their citizens live at or below the poverty level. Commercial nuclear fission also is not safe, says del Callar, particularly for developing nations. They typically receive "old" technology, and they are victimized by US and other outside corporations that have a monopoly on nuclear knowledge and on goods and services. Westinghouse, for example, is building the Philippines' reactor (construction is not yet completed) on a 1972 design. Moreover, Westinghouse argued successfully in a US court, says del Callar, that it was not required to make US-government-mandated safety changes (on the Philippines' plant) after the Three-Mile-Island accident. As a result, less than half of the hundreds of safety requirements in US nuclear facilities will be incorporated in the Philippines' plant.

Despite similar financial problems with commercial nuclear power in his country, Argentina, Carlos Araoz argues in the fifth chapter in favor of nuclear-generated electricity. An official on Argentina's National Commission for Atomic Energy, Araoz is helping his nation acquire commercial nuclear technology. In "Setting the Problem of Nuclear Energy in the Developing World Context," Araoz argues that developing nations have the right to formulate their own energy policies for themselves, without any foreign control. He also argues that, despite its costs, commercial nuclear fission can help his country achieve energy independence and raise per-capita energy consumption. Reducing energy use and criticizing commercial nuclear power are appropriate stances only for industrialized nations, says Araoz. Because per-capita energy consumption in developing nations is only one-fourteenth of what it is in industrialized countries,

Araoz claims that developing nations do not have the luxury of rejecting commercial nuclear fission.

Also representing the view of developing nations, B. C. E. Nwosu likewise supports commercial nuclear power in his essay in chapter six. Nwosu, a Nigerian physicist and technological consultant, argues that atomic energy is economically necessary for citizens of his country, if they are to avoid being fourth-class citizens. In "Issues and Experiences Concerning Nuclear Energy and Nuclear Proliferation," he also discusses the Nuclear Non-Proliferation Treaty of 1968. Like his African countryman, Bena-Silu, Nwoso maintains that the treaty is unfair. Nevertheless, he maintains that no Africans need the bomb. First, they should feed their people and develop their countries, both of which can be accomplished with the help of commercial nuclear fission. According to physicist Nwosu, soft technologies can contribute very little to economic development in the African nations.

Representative of an environmental NGO based in Kenya, Ann Heidenreich is the author of the seventh chapter. In "Future Energy Choices in Light of the Greenhouse Effect," she argues, contrary to Nwosu, that soft technologies like biomass, solar, and small-scale hydro can contribute to African economic development. Introducing the policy issue of whether nuclear energy can help reduce the greenhouse effect, Heidenreich surveys the problem of global warming. A major premise in her argument is that in making energy choices, we should follow "the ecological principle." This principle is that we should use resources close to the same way they are used in natural ecosystems; that is, we should maximize recycling and minimize waste. On the grounds of the ecological principle, Heidenreich argues against use of commercial nuclear fission, largely because of its long-lived hazardous wastes. She also shows that nuclear energy can make only a small contribution to reducing greenhouse gases.

Janos Pasztor, physicist and representive of a Swedish environmental NGO, agrees with Heidenreich that nuclear fission can make very little contribution to the reduction of greenhouse gases. In chapter eight, "What Role Can Nuclear Power Play in Mitigating Global Warming?" Pasztor points out that no new commercial reactors have been ordered in the US since 1974. Although some nuclear proponents have argued that the new "inherently safe reactors" could help solve both energy problems and the greenhouse build-up, Pasztor argues that such new reactors are a long way off. None of them has yet been built, and massive funding would be needed, even if they proved suitable for implementation. Hence Pasztor concludes that the "inherently safe reactors" are unlikely to change the current nuclear picture. The absence of any final repository for high-level waste, public aversion to nuclear power after Chernobyl and Three Mile Island, and the debt crisis in developing countries are all reasons, says Pasztor, why any increase in use of commercial nuclear fission is unlikely. Documenting his claims with extensive and up-to-date studies, Pasztor concludes that energy efficiency in industrialized nations is one of the best ways to address greenhouse problems. Nuclear energy, he claims, could make a contribution in this area that is "only marginal."

Following the historical, technical, and policy discussions in sections one and two of the book, the third and final part of the volume addresses the ethical issues associated with use of commercial nuclear fission. In the ninth chapter, journalist and religious activist Joyce Hollyday introduces some of the ethical issues by recounting the effects of the Three-Mile-Island (TMI) accident on the surrounding community. Surveying the increase in rashes, thyroid problems, leukemias, spontaneous abortions, stillbirths, depression, and fear among the farm families near TMI, Hollyday focuses on nuclear safety. The major ethical issue underlying her personal account is the ability of innocent victims of technology to control their fate, and what their rights should be. The message driven home by her chronicle is that TMI deaths and injuries have been covered up; that accident data have been suppressed; that bureaucrats have been insensitive to the plight of the victims, and that innocent people have lost control of their lives and health. In a compelling, low-key style, Hollyday raises the issue of citizen consent. Are nuclear-accident victims merely examples of unwilling human subjects of involuntary experimentation? Hollyday suggests that they are, despite the heroic efforts of dedicated townpeople to know the truth and to help fellow citizens. She also suggests that, if nuclear generation of electricity continues, then the personal, human, cultural, physical, and psychological toll will be high.

Likewise emphasizing the safety issue, in the tenth chapter Rosalie Bertell addresses the ethical implications of the casualties caused by commercial and military use of nuclear energy. An epidemiologist-biologist and director of an international foundation that studies public health, Bertell argues that current commercial use of fission technology is unethical. It has imposed inequitable burdens on many persons, such as nuclear workers and indigenous peoples who have borne the costs of uranium mining, milling, and waste storage. Bertell also points out that nuclear activities since 1945 have caused millions of casualties, in part because of the unethical and biased ways in which lenient radiation-exposure standards have been imposed on the public. The casualties are not obvious to all persons because the increase in cancer and genetic damage is statistically observable, even though not all victims realize the cause of their ailments. The Chernobyl accident, for example, will cause between 100,000 and 600,000 deaths, according to Bertell, but because of the latency period of disease, not all victims will recognize the source of their illness.

In the eleventh chapter, philosopher and mathematician Kristin Shrader-Frechette addresses many of the ethical issues raised by commercial nuclear fission and by generation of radioactive waste. After providing a revisionist account of the history and economics of nuclear energy, Shrader-Frechette argues that the epistemological and ethical problems associated with nuclear waste suggest that it is unethical to generate electricity by means of atomic energy. In the current situation, radiation risks are distributed inequitably, and citizens have not consented to them. To remedy these problems, Shrader-Frechette proposes that we ensure full compensation and full consent from those affected by nuclear risks, and that we promote use of cost-effective, renewable energy technologies. Shrader-Frechette closes her contribution by arguing that her position on nuclear energy and on renewables is economically defensible, as well as ethically desirable.

Continuing the discussion of ethical issues associated with energy choices, theologian Peter Sedgwick raises the general question of whether moral consensus, on nuclear energy or on any other issue, is possible. In the twelfth and final chapter, Sedgwick argues that we must fight for agreement on policy issues and engage in moral reasoning, in part because we are called to create a community of purpose and value. Our charge is to cooperate in the dynamic activity of God, says Sedgwick, and to transform our environment in wise and sensitive ways. One way to discover whether we are transforming the world in such ways is to apply the directive given in Proverbs 13:22. There the wise person tells us that the good person is one who provides an inheritance for his children's children. One of the ultimate ethical questions for the nuclear debate is whether this technology enables us to provide for our children's children.

2. Moral Consensus and Factual Dissensus

As the preceding survey of the contents of this volume reveals, the chapters agree neither in their points of view nor in their conclusions about nuclear power. We present these essays, in all their diversity, as testimony to our struggling toward conversion; our commitment to democratic decisionmaking; and our efforts at forging a reasoned and religious consensus on the issue of commercial nuclear energy. So far as basic values are concerned, all the essays reveal a similar commitment to important ethical principles. Virtually every author subscribes to the principles, for example, that we we ought to distribute energy equitably, both within and among nations, and that we ought to use commercial nuclear fission in a way that safeguards all the inhabitants of the planet, present and future.

The differences in points of view among the authors of these chapters often arise from factual, rather than ethical, issues. For example, Bena-Silu argues that Chernobyl caused at least 30 deaths, whereas Pasztor claims that the casualties could go as high as 28,000, and Bertell shows that they could be in the hundreds of thousands. Obviously our divergent ethical conclusions are in part a function of our different factual beliefs, and precisely these factual beliefs are at issue, even in the most basic areas. German physicist and churchman Ratsch, for instance, claims in his chapter that "hardly anyone today believes in a nuclear future," while Zairean physicist and churchman Bena-Silu argues that "nuclear energy has come to stay forever." International Atomic Energy Agency spokesperson Linsley claims in his essay that "the associated hazards to humans from such [low-level] waste are not high," whereas university professor Shrader-Frechette and epidemiologist Bertell argue that radwaste risks have been systematically underestimated and have caused thousands of casualties. Likewise Linsley argues that the "most important potential effect of radioactive materials is induction of cancers in humans," whereas Bertell argues, in her essay, that radiation-induced genetic casualties far exceed cancers, and that these casualties have been covered up by industry and government officials. Linsley also argues that "the radioactive impact of waste repositories will be small," whereas Bertell, Pasztor, and Shrader-Frechette argue that, given the necessary containment of some wastes "in perpetuity," there is no assurance whatsoever that the impact will be small.

Even scientists representing the developing world do not agree on whether nuclear energy is desirable for their nations. Bertell, del Callar, Heidenreich, Pasztor, Ratsch, and Shrader-Frechette argue that commercial nuclear fission is not only unsafe but also wildly uneconomical, especially for developing nations, in part because of their debt crisis. Araoz, Bena-Silu, and Nwosu, however, maintain that even non-industrialized nations must pursue commercial nuclear fission, both to achieve energy independence and to allow their citizens to become technologically advanced. Likewise Heidenreich and Pasztor argue that greater efficiency, soft energy, and renewables are desirable alternatives to nuclear power, whereas Nwosu claims that soft technologies cannot make a significant contribution to the energy crisis in developing nations.

At least part of the issue with developing nations' use of commercial nuclear fission is whether they have the right to make the same mistakes, in order to obtain the same benefits, as industrialized countries. Now that the industrialized nations have climbed aboard the lifeboat of economic prosperity, do they have the right to pull up the ladder while developing countries are still in the water? Are dangerous, but beneficial, technologies like ladders that developing nations need to use? Or, do industrialized countries have the right to impose a double standard on the rest of the world and to behave in paternalistic ways? This issue of paternalism, in turn, raises a more fundamental ethical question. Does the right to self determination - for example, in energy policy - include the right to be wrong?

The purpose of our presenting these essays, with all their factual and ethical differences, is to invite the entire human community to investigate and evaluate our divergent opinions and to share our ongoing discussion. We ask you to debate the issues as presented here, and to help us to forge a reasoned and religious consensus. Our hope is the same one that John Kennedy expressed in 1963, in a university address:

> *Let us not be blind to our differences - but let us also direct attention to our common interests and the means by which those differences can be resolved.*

SECTION ONE

Historical Survey of Commercial Nuclear Technology and Responses of the Churches

An Overview of the Issues Confronting Nuclear Energy and the Concerns of Churches

Ulrich Ratsch
Forschungsstatte der Ev. Studiengemeinschaft
Heidelberg, Federal Republic of Germany

Energy, and especially nuclear energy, has been on the agenda of *Church and Society Sub-Unit* of the WCC at least since 1974. The debate reached a peak around the middle of the seventies and lasted until about 1983. The energy issue, of course, was never entirely abandoned. But after that date, the Church and Society Sub-Unit was mainly concerned with other topics, as can be seen from the sub-unit's reports of publications.[1] Focusing attention on other topics signifies neither acceptance nor rejection of nuclear power or any other specific option of energy policy. Also, it must not be understood as indicating that the energy issue had become unimportant.

In the following section I shall try to highlight the major events and the focal points in the ecumenical debate on nuclear energy. I shall largely quote from official WCC documents to give a picture as authentic as possible.

1. Stations of the Debate

The phase of the most concentrated discussions and the densest succession of meetings started in June 1975 with the "ecumenical hearing on nuclear energy" in Sigtuna, Sweden. This meeting is documented in the publication *Facing Up to Nuclear Power*.[2] At the time of this meeting, the public debate on energy issues was highly polarized and controversial. "It was therefore important that in entering this particular debate the World Council of Churches should try to introduce a note of fairness. At Sigtuna there was consequently a wide range of positions, including those for and against and some people hovering around the mid-point of positions in between. The achievement of Sigtuna was to reflect fairly these divergences. It is not surprising that in the continuing public debate these differences persist and that they should be reflected in the evaluation of the ecumenical report."[3] A lot of individuals or church institutions within member churches of the World Council took up the points raised at Sigtuna and introduced them into their respective national debate. There were a number of favorable reactions to the Sigtuna Report, praising the clarity and fairness in documenting the various opinions on all the relevant issues in the nuclear debate. There have also been comments to the effect that this report was "equivocal" or too "neutral." The division of Church and Society of the National Council of Churches in the US, for instance, stated: "While we respect the World Council's choice of process it was the division's conviction that, since people are being asked to make decisions for or against nuclear energy today, to remain 'neutral' is in practice to support continued growth of nuclear capacity in the US and increasing reliance upon it. World Council's 'balanced' report failed to arouse debate on the issue within the US church community."[4]

2. Fifth Assembly of the WCC, Nairobi, November/December 1975

"The fifth assembly of the World Council of Churches' meeting in Nairobi, Kenya, November/December 1975, received the Sigtuna Report on nuclear energy and in turn transmitted the following statements (and recommendations) to the churches:

> "Nuclear energy. Increasing pressures on the world petroleum market bring the prospects of rapid expansion in the construction of nuclear power stations in countries at widely differing stages of development. This poses ethical dilemmas in view of (1) the inevitable coupling of nuclear energy for electricity generation and the production of nuclear weapons: how to limit the use of nuclear energy only for peaceful purposes of development? (2) the hazards involved in the storage of nuclear waste for long periods; and (3) the problems of theft and sabotage. The ecumenical discussion of these issues has just begun and needs to be continued. Discussion should concentrate on the effectiveness of the monitoring and control of the nuclear fuel cycle at the international level, ratification of the Non-Proliferation Treaty by all countries possessing nuclear technology, the drift towards a plutonium fuel economy and its consequences, and the political and military implications of the further spread of nuclear technology.
>
> Alternative Energy Technologies. During the next decade, there will be increased research and development of alternative systems of energy production. The world is now entering the transition between dependence upon petroleum to dependence upon other sources of energy, e.g. solar, wind, water and geothermal. Priority must be given to ensure that research on alternative energy sources is funded at a high level, especially as these technologies will be of direct benefit to developing countries."

These statements were then supported by the following recommendation to the churches:

> "Nuclear Energy. As far as nuclear energy is concerned the churches at the national level should lobby for more effective safeguards in the control and operation of the nuclear fuel cycle to ensure that there is a reduced risk of diversion of strategic weapons materials; in addition there is a need to press for more serious investigation of the long-term disposal of radio-active wastes by those responsible for the management of nuclear power programs. The churches are further urged to register their reactions to the findings of the Ecumenical Hearing on Nuclear Energy, Sigtuna, Sweden, 1975. (See "Report on Nuclear Energy," Anticipation No. 21, W.C.C., October 1975.)

Alternative Energy Technologies. Since there are inevitable grave ethical problems involved in the application of nuclear energy, we recommend that the churches and individual Christians seek to promote sufficient funding and moral support for research and development of renewable sources of energy, e.g., solar and wave energy."[5]

3. Bucharest, Rumania (1974)

In 1974 the World Council of Churches convened a conference, entitled "Science and Technology for Human Development," in Bucharest, Rumania. It was the result of five years of preparatory work within the WCC study program on the future of persons and society in a world of science-based technology.[6] "At this meeting, the churches were inevitably drawn into the growing controversy concerning the different energy options available, and especially concerning nuclear energy."[7] Five major areas of concern were defined which, since then, have characterized the ecumenical debate on energy issues:

- ethical components of energy options;
- the importance of wide public debates;
- the need to set up an "international resource bank" to help commercial energy development in developing countries;
- the need for energy conservation, more rational energy use in those countries where too much is used; and
- the need to develop alternative energy sources based onsolar and other renewable forms.

4. The Glion Consultation (1976)

As a result of the discussions in Bucharest (1974), Sigtuna (1975), and Nairobi (1975), it became definitely clear that nuclear energy could not be discussed in isolation. Rather, it had to be discussed "in the context of looking at all the energy options available."[8] The Working Committee of Church and Society invited several internationally renowned energy consultants to its annual meeting in Glion, May 1976. Their discussions produced the idea of "energy for a just and sustainable society." "At the Glion meeting, the energy crisis in the developing countries emerged as a key issue. The group became more conscious of the extreme seriousness of the energy situation in many developing countries, where dwindling supplies of firewood, increased populations, and prohibitive prices for petroleum and other commercial energy products...were not only creating inconveniences (as in the industrialized countries) but resulting in the inability of many people to even cook their daily food requirements or to irrigate their lands."[9]

"As a result, the Glion meeting mandated the Church and Society Sub-Unit to create the "Energy for My Neighbour" program (ENP). The idea of this program was, to put it very briefly,

- to study the special energy problems of Third-World societies;
- to give special attention to the views of Third-World populations on these problems;
- to promote energy conservation in industrialized countries; and
- to use the money saved by these measures to finance studies and energy projects in the poor regions of the world.

Also, the Glion consultation recommended a continuation of the energy debate and, in particular, further discussions on the ethical, social, economical and ecological problems related to the use of nuclear technology. This work, so it was decided, was to include as an integral part "investigations into the other alternatives for possible energy supply especially the solar options."[10]

5. The IAEA conference, Salzburg (1977)

In May 1977, the International Atomic Energy Agency (IAEA) convened an "International Conference on Nuclear Power and Its Fuel Cycle" at Salzburg, Austria. The WCC was officially invited to submit a paper on the ethical issues and the public acceptance of nuclear power. It became one of three presentations in a plenary session on "Nuclear Power and Public Opinion." The paper was prepared by members of the energy advisory group of the Church and Society Sub-Unit, and it was endorsed by the central committee. It stressed five major points.

1. Nuclear energy offers the potential of providing a large part of the world's energy needs, counterbalanced by some exceptional risks.
2. The maturity of the nuclear energy system is not yet such as to justify large scale, or unlimited, applications.
3. The right of access of all nations to nuclear technology must be recognized.
4. The issues of access and security involve the collective responsibility of all nations; they need to be assessed anew and administered through the IAEA.
5. Public confidence in the use of nuclear energy, seriously shaken in recent years, can be revived only by the widest possible public discussion of the technical options and of the value judgments underlying present patterns of energy consumption."[11]

6. Bossey (1978)

This consultation was "convened at the ecumenical institute in Bossey, "...with the participation of 60 energy experts, church leaders, and theologians, representing a wide variety of opinion, including those strongly opposed and those substantially in favor of nuclear energy."[12] Discussions at this meeting became often as polarized as the secular debate was. No attempt was made to reach a consensus or to compromise the views of any participants. The main ideas developed at this consultation were as follows:

- "The nuclear debate cannot be addressed...without consideration of other energy options, including especially energy conservation and the possibil ity of technologically less complicated societies.
- Energy consumption and more rational use of energy deserve much more attention.
- The nuclear debate is but symptomatic of a much deeper societal debate about more technology or less, about centralized versus decentralized....with many groups defining these terms to suit their own social aims.
- No grounds exist for *rejecting* nuclear power categorically. On the other hand, it cannot be *accepted* categorically; it is a conditional good, subject to reasoned acceptance under some circumstances and subject to reasoned rejection in others. Even where accepted there may be a point when we may have to say 'Thus far and no further.'"[13]

Again, many people found these statements useful, whereas others, both pro and contra nuclear energy, rejected them.

On all the occasions mentioned, the moratorium issue had been discussed. In the ecumenical debate, until the beginning of 1979, there never was a clear majority for a moratorium on the issue of nuclear energy. So the proposal of it never showed up in the final statements. This changed in 1979. In March of this year the largest accident in the published history of civilian nuclear programmes, so far, occurred at Three Mile Island, Harrisburg, Pennsylvania. It came close to a catastrophic accident and it showed, in the opinion of many people, that highly complex technologies like nuclear energy can never be made entirely failsafe.

7. The "MIT-Conference" (1979)

Consequently, at the "Conference on Faith, Science and the Future," at MIT, Cambridge, Massachusetts, July 1979, "not only for the energy section but perhaps for the whole conference the nuclear energy question became the most popular item."[14] In long and heated debates in the energy section of that conference a long list of guidelines to member churches and to national governments was formulated. "During the plenary debate of the report of the energy section a motion for a nuclear moratorium was introduced. After long and heated debates the moratorium was passed by a three to one majority and was inserted as one of the guidelines."[15] The guidelines include these recommendations to the churches:

- to promote wide, open public debates in which all sectors of the society should be enabled to participate;
- to identify ethical criteria by which the social impacts of (energy) technologies must be assessed;
- to develop a climate of trust and critical self-awareness (within these public discussions) in which biases can be acknowledged and corrected;

- to engage in constructive and critical analyses of the national and international energy policies; and
- to endorse, promote, and support the use of the WCC "Energy for My Neighbour" program.

They also include recommendations to governments:

- to promote studies on energy use and demand-oriented energy forecasts;
- to promote energy conservation and more rational use of energy;
- to promote the use of all sorts of renewable energy;
- to limit the growth of the combustion of fossil fuels;
- to introduce immediately a moratorium on the construction of all nuclear power plants for a period of five years;
- to devise energy policies on a national and international level in a way that achieves more justice and participation of all people.

It is necessary to stress here that nuclear energy was just one of many topics to be discussed at the MIT meeting. All other forms of energy were investigated with their potential, their economic costs, their ecological or societal hazards. In particular, all forms of renewable energy were examined. The special energy problems of developing countries ("the other energy crisis") have always been given much attention. Despite this, the discussions in the ecumenical circles until the MIT Conference have, at least in part, reflected the public debate in western, industrialized countries and have, at least in part, been dominated by people representing this debate.

8. Third-World Conferences (1980-1982)

Under the auspices of the "Energy for My Neighbour" program in the years 1980 to 1982, a succession of three conferences was held in Third-World countries (Madras, India; Chiang Mai, Thailand; Lima, Peru). Participants in these regional consultations included technical experts in energy development, community organizers, government representatives, church members and leaders, as well as representatives of the Roman Catholic churches and of other faiths, such as Muslims, Hindus and Buddhists. They all came from Third-World countries. They predominantly expressed the view of the local people.

The guiding scheme of all these consultations was "towards more just participatory and sustainable energy policies for development." Low cost, simple energy technologies relying on locally available resources have been investigated with their advantages and disadvantages. Nuclear energy in these consultations has hardly been mentioned. It was not deliberately omitted from the agenda by the organizers. It simply wasn't addressed because most of the participants agreed that nuclear energy is not appropriate to relieve the lot of the mass of the people. Extremely densely populated areas in Southeast Asia (e.g. Java) were seen as the only possible exception.

9. Topics Addressed During Eight Years of Ecumenical Debate

The ecumenical debate inevitably reflected the public discussions in North America, Western Europe, Latin America (particularly Brazil) and some Asian countries (for instance, India, Indonesia, Malaysia, and the Philippines). Hence, the same focal points of controversy were found. They include the following questions.

- What should be the global and national distribution of energy resources and energy demand?
    - What are the advantages and disadvantages of various energy sources (fossil fuels, hydropower, nuclear electricity, direct and indirect forms of solar energy, especially biomass)?
    - Do we correctly distinguish the two energy crises, the oil price rise and the nonavailability of fuelwood? How are they related?
    - What can be the role of energy conservation with respect to a more rational use of energy?
    - What type of lifestyle belongs to "soft" or "hard" energy paths, centralized or decentralized energy supply systems?
    - What are the social and political implications of different energy paths?
    - What is the effect of different energy paths on discrimination, both on an international and national level? Are we promoting energy for the rich or for the poor?

- How does energy supply play an integral part of economic and social development?
    - What do we mean by "sustainable development," as postulated by many churches and other groups in the ecumenical debate?
    - Which people and which sectors of society benefit most from different energy technologies?
    - What are the energy sources needed in rural areas, on the one hand, or in the industrial sector, on the other? Is it possible to finance the promotion of all these energy technologies?

- What are the ecological problems related to the use of energy?
    - What is the impact of various energy sources on deforestation?
    - How do we assess the pollution problems associated with different energy systems (e.g., acid rain, the greenhouse effect, the displacement of food crops)?

- In what ways is nuclear energy beneficial or harmful?
    - Can nuclear technology be safely managed, or does it pose an unacceptable risk of radioactive pollution to the population?
    - What are the hazards involved in the storage of nuclear waste for long periods, and can a solution be envisaged for this problem?

- How do we judge the ethical dilemmas, in view of the inevitable coupling of nuclear energy for electricity generation and the production of nuclear weapons?
- Is there really a significant risk of theft and sabotage? If so, then how can we cope with it?
- Is nuclear energy a long-term option without the introduction of fast-breeder reactors? Do breeder reactors significantly change the risks and hazards mentioned above?
- Is nuclear energy economically feasible? If so, thenfor which countries?
- Does the proliferation risk necessarily lead to an increase of discrimination between nuclear "haves" and "have-nots"?
- Does the use of nuclear energy diminish or, on the contrary, enhance existing social justice, both on a national and on the local scale? In particular, is nuclear energy an appropriate technology to ease the lot of the mass of the people in the so-called developing world?

And finally:

- What are the alternatives to the introduction and use of nuclear technology?

As the WCC documents show, and as I know from my own involvement in the ecumenical debate, nuclear energy has never been discussed in isolation. It has always been looked at as part of an entire energy system. The problems of fossil fuels have never been neglected, and the uncertainties about the potential role of renewable energy sources, particularly solar energy, have clearly been faced.

It may be interesting to note that the Working Group of Church and Society, together with a group of energy consultants, in the report from their meeting at Glion, as early as May 1976 says: "The carbon dioxide level in the atmosphere (from burning fossil fuels) will have risen far enough to bring us to the brink of substantial climatic changes of uncertain nature."[16]

10. Are There Any Entirely New Topics?

Actually, this paragraph should start with the question: "Have any of the issues mentioned earlier become obsolete?" It seems obvious to me that all of the listed topics are still relevant. Of course, figures of supply and demand have changed; the knowledge about ecological hazards has been widened; the risk assessments for various technologies have been refined. Yet, the overall picture and the overall evaluation of costs and benefits, advantages and disadvantages related to the different energy sources, have remained the same. "Costs and benefits" here are not only to be understood in financial terms. But looking at financial costs in particular, the advantage of nuclear power has diminished, if not vanished. The result of cost comparisons varies from country to country, depending on technological development and resource endowment. Official calculations of electricity and generating costs indicate an advantage for coal in some countries, for nuclear in others.[17] Scientists from some developing countries, e.g., Brazil or India, claim that nuclear electricity in their countries is cheaper than electricity from any other available source.[18] Yet, looking at the overall economic burden for their societies, most developing countries estimate nuclear energy to be too expensive. This is not a new or recent insight; they drew the same conclusion 10 years ago.

Global climatic change and deforestation on a massive scale have now been recognized by a broad public, but they have been discussed since 1976. Many of the threats and problems were already seen at that time.

11. So What is New?

With regard to nuclear energy, there is the hope that small-scale, cheap nuclear power plants can be built (e.g. the "Module-HTR"). Both India and Argentina started programs to develop "small or medium power reactors" (SMPRs), the Argentinean program being stalled at present due to financial problems.[19] Other developing countries hope that supplier countries in the north will construct SMPRs, as they have promised. This could open new markets for the nuclear export industry. Small power reactors could help supply electricity to consumers in far more countries than nowadays could afford it. But even with the small power reactors, electricity will be the most expensive form of energy. The question remains: who are these consumers who can afford electrical appliances?

Generally, growth of energy demand, particularly demand for electricity is less than expected in the seventies.[20] Also, oil prices at present are relatively low. Even if prices rise again, the increase is expected to be less steep than in the seventies. Both these facts mean that today we can discuss energy futures under less urgent pressure than we felt in the seventies. These facts, in addition, were part of the reasons for (what some people call) a "de facto moratorium" on nuclear energy. Construction of nuclear power plants has come to a virtual standstill in most First-World countries.[21] The exception is France. But there also, the programs have been cut down and some power plants already have been shut down because of a tremendous overcapacity. The main reason for the cessation of the nuclear programs that were devised and debated in the seventies are economic problems together with revised forecasts of electricity consumption. Only the so-called Second-World nuclear programs are nearly unchanged. But even there, construction is slowed (construction stoppage in Poland and Yugoslavia, abandoned construction plants in the USSR due to ecological considerations). Hardly anyone today believes in a nuclear future, i.e., a future dominated by nuclear energy.

Another new factor is the experience of the Chernobyl accident. Millions of people have experienced the consequences of a major nuclear accident, ranging from permanent evacuation from their homes to minor restrictions in their daily diet.

The focus of interest, particularly in Third-World countries, has shifted from nuclear-power production to non-energy uses of nuclear technology in medicine (irradiation of cancers, radioisotopes as tracers in diagnosis) and in agriculture and food processing (irradiation for the purpose of conservation of food).

With regard to other energy options, there is nowadays a greater awareness of the environmental problems of all known energy sources. As a result, the potential of renewable energy sources (solar, wind, geothermal, biomass and others) is assessed in a more sober way, and with less enthusiasm.

Public awareness of the necessity to conserve energy has increased considerably since the mid-seventies. Also, there have been significant achievements. But, on the other hand, the recent past is characterized by complacency and decreasing interest in energy saving. In particular, government policies to support energy conservation in many western countries are far from being satisfactory.[22] Likewise, research and development efforts in the fields of renewable energies are not sufficient and fall far behind official announcements.

12. Expectations with Regard to the Future

The WCC conference in Kinshasa, Zaire, October 1989, was entitled "Consultation on Nuclear Energy". If at present we limit our discussions to *nuclear* energy alone, we should fall behind the level of discussion reached in 1975. I strongly hope that we shall be able to avoid this mistake. Nuclear energy has to be discussed as one option among others in the energy system.

In discussing energy systems we should distinguish between long-term options (e.g., renewable sources of energy, changing lifestyles in the highly industrialized countries, ecologically sound ways of using biomass energy) and short-term necessities (meet the urgent needs of the majority in the developing countries, perhaps by increasing the use of fossil fuels in the near future; provide a sufficient supply of firewood while trying to avoid too serious ecological damage; try to devise agro-forestry schemes aimed at providing food, fuelwood, and fodder, while at the same time conserving soil fertility).

The churches, in their debate on energy problems, always have been particularly concerned with the situation of the mass of the poor people. Their emphasis on the questions must be: what type of development is likely to improve the situation of the poor? Appropriate energy systems must be assessed with regard to their suitability for this development path. Sustainable development can take place only when the natural resource base is maintained, or better, improved. Therefore the ecological consequences of energy systems have to be assessed, both with respect to the short term and with respect to the long term. Sustainable development can be achieved only through a participatory process. And if so, then we must ask which energy systems lend themselves to self-determined development and use by the poor.

Notes

[1] Cf. the minutes of the Working Committee meetings of the Church and Society Sub-Unit, WCC; the series *Church and Society Documents* or: Howard Davis and David Gosling (eds.), *Will the Future Work*, Geneva, WCC, 1985.

[2] John Francis and Paul Albrecht (eds.), Edinborough, Andrew Press, 1976.

[3] *Anticipation 23*, Geneva, WCC, November 1976, pp. 9-10.

[4,5] *Anticipation 23*, Geneva, WCC, November 1976, p.11.

[6, 7, 8, 9, 10, 11, 12, 13, 14, 15] *Anticipation 29*, Geneva, WCC, November 1982, p. 26.

[16] *Anticipation 23*, Geneva, WCC, November 1976, p.2.

[17] OECD: Nuclear Energy Agency, *Nuclear Energy in Perspective*, Paris, OECD, 1989, pp.22-23.

[18] Cf. M.R. Srinivasan, "The Economic Case for Small and Medium Power Reactors in India," in *Small and Medium Power Reactors 1985*, ed. by IAEA, Wien, IAEA, 1986, pp.15-30; compare also the contribution of Carlos Araoz in this volume.

[19] A short description of the program is given in "CAREM: Un reactor modular de baja potencia," *Revista Argentina Nuclear* 16 (December 1988): 56-57.

[20] Cf. for instance, OECD: Nuclear Energy Agency and the International Atomic Energy Agency (eds.), *Nuclear Energy and its Fuel Cycle, Prospects to 2025*, Paris, OECD, p.19; OECD/NEA (eds.), *Nuclear Energy in Perspective*, Paris, OECD, 1989, pp.17-18.

[21] OECD/NEA (eds.), *Nuclear Energy in Perspective*, Paris, OECD, 1989, p.17.

[22] OECD (ed.), *Energy Conservation in IAEA Countries*, Paris, OECD, pp.7-15, 19-20, 43.

Radioactive Wastes and Their Disposal

Gordon S. Linsley
International Atomic Energy Agency
Vienna, Austria

1. Introduction

The radioactive waste issue is widely regarded as being one of the most important in ongoing nuclear debate. In some countries the future of the nuclear power industry has been made conditional upon its satisfactory resolution.

The issue is not a purely technical one, and much of the opposition to radioactive waste disposal has its base in the emotional response of people nowadays to any issue involving radiation or radioactivity. There are also issues of equity involved, expressed in the resistance of local population groups to accepting the risks from radioactive waste by having disposal facilities located near to where they live (the so-called "not in my back yard" or NIMBY syndrome). The notion that the radioactive waste disposal problem is "unsolved" from a technical standpoint also has credence with many people and, for example, the recent report of the World Commission on Environment and Development contains such a statement.[1] This is not the view of experts in the field of radioactive waste management,[2] and it appears to ignore the real and concrete developments in several countries.[3,4,5]

From the beginning, plans for the final disposal of radioactive wastes have raised new and challenging issues for society largely because of the long time scales over which safety has to be assured for some types of waste. Other types of hazardous waste, notably those containing heavy metals, present the same problems; they are highly toxic and must be isolated from humans forever. But this does not seem to have been widely recognized. In reality, because of the larger waste volumes involved, they present at least as serious a problem for society as radioactive wastes.[6,7]

It is important to be aware that, because of the widespread use of radioisotopes in medicine and research, there are radioactive wastes which must be safely managed in most countries in the world. Obviously, the quantities for disposal are very much greater in countries with nuclear power programs, but many aspects of the approaches which must be used for their management are similar.

The aim of this review chapter is to describe the current international status of radioactive waste disposal. In the following sections there are discussions of the various disposal options for the different types of radioactive wastes; the development of safety criteria; the methodologies and results of safety analyses; and, finally, comparisons with the approaches to the disposal of other types of wastes.

2. Sources and Characteristics of Radioactive Wastes

In the nuclear fuel cycle, radioactive wastes arise in the mining and milling of uranium ores, the enrichment process, the fuel-fabrication plants, the operation of nuclear reactors, the reprocessing of spent fuel, and the decommissioning of nuclear facilities. Wastes also arise from the use of radioactive substances in hospitals and research institutions, although on a smaller scale.

The hazard presented to humans by radioactive wastes arises from the radionuclides contained in them. The emission of ionizing radiation from the radionuclides is potentially harmful. Exposure to ionizing radiation may occur as a result of the intake of radionuclides by inhalation, if they are in the form of a gas or an airborne particulate, or by ingestion if they are present in food or drinking water. In these respects they have the same intake and exposure routes as many other hazardous substances. Only the additional route, that of direct exposure of the body to external radiation, is unique to radionuclides. The potential consequences of exposure to radioactive materials can also be compared to those from other hazardous materials. For many of these substances and for radioactive materials, the most important potential effect is the induction of carcinogenesis in humans.

Radioactive wastes can be placed in three broad categories according to their activity content: low-level wastes, intermediate-level wastes, and high-level wastes. Low-level wastes consist of items such as paper, clothing, and laboratory equipment which have been used in areas where radioactive materials are handled, contaminated soils, and building rubble. Intermediate-level wastes include ion-exchange materials which are used in the treatment of gases and liquids before they are discharged into the environment. The term 'high-level waste' is used to refer to the highly active liquid which is produced when spent fuel is reprocessed or to the spent fuel itself when it is regarded as high-level waste. Table 1 shows the quantity of radioactive, as compared with non-radioactive, wastes; it indicates that the volumes of low-level waste far exceed those of the other radioactive-waste types.[8] Although the term, 'low-level radioactive waste', is not precise, and a large variety of waste types from different origins and with different radioactive contents fit within the description, it can be generally stated that the associated hazards to humans from such wastes are not high; the hazards are chronic in nature rather than acute and comparable with the risks from many common types of waste contaminated with chemicals.[9,10]

From a disposal point of view, it is of more interest to classify the wastes according to their dangerous life-length. For the radionuclides that normally appear in the radioactive waste from nuclear-energy production one can distinguish between two types.

- Short-lived waste: The radioactivity has decayed to innocuous levels in less than 300-500 years.
- Long-lived waste: The radioactivity remains important for hundreds of thousands of years.

The waste in the second category normally contains substantial amounts of transuranic elements, such as plutonium and neptunium, whereas waste classified as short-lived mainly contains activation products, such as cobalt-60 and fission products, such as caesium-137. From nuclear-power production, the spent fuel and wastes from reprocessing of the spent fuel are long-lived, while most of the wastes from reactor operation are short-lived.

3. Waste Management

The objectives of waste management are to process the wastes in such a way that they are in a suitable form for storage and disposal, and to dispose of them so that there are no unacceptable risks to present and future generations of humans. Low-level wastes do not generally need processing; they can be packaged, sometimes after compaction, and disposed of directly. Most intermediate-level wastes do not arise in a form which is suitable for direct disposal; they have to be incorporated in an inert material such as concrete, bitumen, or resins prior to disposal.

High-level wastes (HLW) from reprocessing are in liquid form, and the currently most favored approach for their immobilization is to incorporate them into a glassy matrix. Facilities exist or are being developed for this purpose in several countries. Views vary on the need for storage prior to disposal of high-level wastes, whether in the form of glass or spent fuel rods, but it is generally agreed that they must, in any case, be stored for a period (at least about ten years). This is to allow heat decay which will facilitate handling and to limit thermal effects on the chosen disposal medium.

4. Basic Principles of Disposal

The two fundamental options available for disposal of any material are either to endeavor to keep it in the same place for as long as necessary, or to allow natural processes to mobilize and disperse it harmlessly. The first concept is known as containment (or "concentrate and confine") and is clearly very difficult to ensure for very long periods of time. The second is commonly referred to as "dilute and disperse." There are obvious attractions to the latter concept for materials which remain toxic for extremely long times. Containment of such materials means that they always remain a potential hazard at one place since their concentration is only slowly reduced by decay. It would clearly be better to design a disposal system which allowed for their mobilization and transport by natural processes, resulting in their dilution in a widely dispersed form at concentration levels which presented no unacceptable hazard.

The attractions of "dilute and disperse" are counter-balanced by some very difficult problems. For example, it must be shown that a sufficient understanding of the dispersal of the released radionuclides in the environment has been obtained in order to be able to demonstrate that they present very low risks to humans. It is also necessary to reach agreement as to what constitutes an acceptably low risk. Finally, for releases of radioactive materials, especially in gaseous or liquid form, widespread dispersion may occur, and persons in other regions or countries may be exposed to risks for which they have received no benefit (for example, in terms of energy from nuclear power) and to which they did not agree in advance. These equity issues are not unique to radioactive waste disposal but exist for many practices involving the release of pollutants into the environment.

Both containment and dispersion concepts are included in present waste-disposal systems. For example, short-lived waste might be earmarked for shallow-land burial where total containment may be required for around 300 years. After this, the total concentration of both short- and long-lived radionuclides in these low-level wastes is very low, and almost any degree of dispersal may be acceptable; hence the choice of a shallow, rather than a deep, disposal environment. In the case of HLW the initial containment period may vary from around 1,000 years, up to a million years or more, depending on the concept.

The basic principles of disposal thus involve the selection of a disposal environment and a system of engineered emplacement of the wastes to provide the required containment and/or dispersal. A general requirement for disposal of any waste is to ensure that the disposal location is remote from the human environment; thus the two methods which have always been chosen historically are burial on land or dumping in the sea. Consequently, these two methods have emerged as the principal contenders for disposal of radioactive wastes. It is worth bearing in mind that there have been no adequate controls on disposal of most industrial wastes until very recently (the early 1970s in most countries). This includes the early disposal of radioactive wastes.

Moving back to a more technical summary of the functional requirements of any disposal system, the following list highlights the main points to be considered.

- The time periods over which isolation is required are extremely long (indicating dependence on natural as well as man-made barriers).
- Assurance is needed of low release rates once the period of complete isolation is past.
- Wastes must be removed from the effects of human activities or of catastrophic natural events.
- The technology to implement disposal should be available, and the task should be achievable at reasonable cost.
- It must be possible to model adequately the processes which control the long-term performance of the chosen system, be they physical, chemical, or even biological.

How these requirements can be met by various disposal options is discussed briefly in subsequent paragraphs, beginning with geological disposal.

4.1 Geological Disposal[11]

Geological formations are characterized by their great age and, in most environments, by their considerable long-term stability. The lengths of time over which certain rock formations have retained their structure, integrity, and properties are beyond the scope of human experience; and although the rocks are subject to constant change, the rates are generally extremely slow.

Burial of wastes in a suitable geological formation removes them from the environment and puts them out of reach of major disruptive processes, both natural and human-caused. It also provides more-than-adequate shielding from the radiation emitted by the waste. The wastes are put in an environment where all natural processes which can affect their behavior are very slow. The deeper the burial, generally speaking, the slower and less pronounced are these processes. It is clear that long-lived wastes will generally require deep disposal, but burial at shallower depths may be acceptable for shorter-lived wastes. Political expediency also plays a role in the choice of disposal depth. Thus some countries have chosen deep disposal for even low-activity wastes because they consider that the extra costs involved are more than matched by the political gain of demonstrable over-design of safety barriers.

The natural geological processes which will affect the behavior of buried wastes are dominated by the behavior of groundwater in the host rock unit and in under- and overlying formations. All rocks below the water table are fully saturated with water; that is, their pore spaces, which can vary from a few, to several tens of, percents of the rock volume, are full of water. Even above the water table a considerable proportion of the pore space is water filled. This groundwater moves slowly through the rock and, in doing so, it can leach and dissolve the waste and its containers and carry radionuclides away and potentially back to the biosphere. The behavior of groundwater in the rock (hydrogeology) is thus a very important factor in selecting a host geological environment for disposal. Very slow and predictable rates of water movement are desirable for the environment of the waste repository. Many geological environments fit this requirement and a variety of suitable rock types are available for construction of a disposal system, either deep or shallow.

The disposal systems themselves fall into four main types:

- shallow burial in trenches, tumuli, tunnels, concrete bunkers or caissons;
- deep injection of liquid wastes into porous strata (with or without a slowly solidifying additive such as cement);
- deep burial in a purpose-built or modified existing mine, usually called a "repository";
- deep burial in boreholes drilled from the surface.

At present, the main technique commonly practiced is shallow burial of low-level wastes, although some deep injection of low- and intermediate-level wastes occurs to a limited extent. While disposal of more active waste in unconditioned form took place in some countries in the 1940s and 50s (mainly associated with nuclear-weapons programs), the limited information available indicates that such practices have been discontinued.

Geological disposal is considered to meet the requirements listed earlier. In common with other disposal options the main task is in demonstrating that the final requirement of predictability is met.[12,13,14]

4.2 Alternative Disposal Options

Although geological disposal of radioactive wastes is generally favored, a number of other options have been studied for long-lived wastes. Generally speaking the high potential costs of these other methods have meant that, with the exception of sub-seabed disposal, they have not been considered seriously. The alternatives for long-lived wastes include:

- space disposal;
- ice-sheet disposal;
- disposal beneath the seabed;
- rock melting;
- nuclear "incineration."

Space disposal is one of the most commonly suggested solutions for nuclear waste offered by the "man in the street." It has the great apparent attraction of removing waste from the biosphere for all time. Studies have been carried out by NASA which show that the launch of nuclear payloads into a solar orbit halfway between the Earth and Venus is feasible, based on space-shuttle technology. The main problems involved are the extremely high cost of this option and the risk of launch failure. The disastrous loss of the US shuttle-craft "Challenger" in January 1986 highlighted this risk and reinforced the previously estimated rocket-launch failure rate of 6 in 100 dramatically. The cost alone means that this option could only be considered for the very highest specific-activity wastes, while the launch risks have all but eliminated this disposal route from further consideration. A launch failure could cause widespread pollution to almost any region of the world. The repercussions of the $8 million search and clean-up exercise, carried out in the Arctic winter of northern Canada in 1978, following the re-entry of a satellite nuclear power-pack and its break up in the atmosphere, will long be remembered.

Ice-sheet disposal is another option with great popular appeal. Apart from the remoteness of the polar ice-caps, the fact that heat-emitting wastes would be self-sinking in the ice has clear attractions. The problem areas are the very high transport and handling costs, uncertainties in the evolution of these areas over the geological time spans for which containment must be considered, and legal constraints. At present, the controlling limitation is probably the latter, as waste disposal in the Antarctic is expressly forbidden by international law, while alternative sites in Greenland are controlled by Denmark.

Disposal of high-level, and some types of intermediate-level, wastes on the bed of the ocean is prohibited by international agreement (the London Dumping Convention), although it can be shown that the disposal of some wastes by this method presents very low individual risks.[15] In recent years considerable research has been carried out on disposal beneath the ocean bed. Advantages of sub-seabed disposal in the deep oceans are physical remoteness, chemical and geological stability of the sediments and rocks, and the great dilution capacity of the sea in the event of releases occurring. In addition, disposal techniques involving self-burying containers ("penetrators"), which are allowed to free-fall from the disposal ship, could be relatively inexpensive. Technical problems mainly involve ensuring the performance of simple disposal techniques in the deep waters considered, or the high cost and engineering problems with more sophisticated alternatives involving, for example, emplacement in holes drilled in the ocean floor. Potentially more critical, however, are international concerns. In the foreseeable future, pressure from non-nuclear countries with marine seaboards is likely to produce further restrictions on the use of the oceans for waste disposal. Research and development programs into sea disposal of long-lived wastes are being reduced, not for technical reasons, but because political factors seem likely to rule out these disposal options.

A more speculative alternative to any of the geological emplacement techniques for high-level waste is "rock-melting." This involves emplacement of highly concentrated solid or liquid HLW into a deep borehole or cavity. The radiogenic heat produced by the waste melts both the waste itself and the surrounding rock and, upon eventual cooling, the wastes are incorporated into a natural rock matrix. As yet, the technical knowledge required to evaluate fully waste/rock interactions in this system does not exist, and hence it has been considered with only very low priority.

The final technique considered - nuclear transmutation or "incineration" - is not a disposal option as such, but provides a method for destroying the long-lived radionuclides which present most of the problems in other disposal systems. Very simply, long-lived radionuclides (mainly actinides) are chemically extracted from the waste and emplaced in the neutron flux in a reactor core. This causes them to be activated or fissioned to form shorter-lived products. This technique can effectively destroy many of the most potentially harmful nuclides, but is very expensive with presently available technology. In addition, the large amount of processing involves handling vast quantities of very short-lived, and hence extremely active, radionuclides. Because this work is likely to result in considerable radiation exposure to the work force involved, it must be balanced against potential exposures to future populations, exposures that are predicted to be very low in direct disposal techniques. This is an important factor which also has to be taken into account in waste conditioning. Present exposures to an existing work force are real, whereas predicted small-scale benefits gained in the very long term, for example, by having a more durable waste form, are speculative and lack reality.

5. Safety Principles

The two basic objectives of safe waste disposal are:

- to protect humans from the potentially harmful effects of radioactive waste;
- to dispose of the waste in such a way that no commitments or constraints due to present-day waste are transferred to future generations.

Much work has been devoted to translate these objectives into a coherent set of regulatory requirements such as safety principles, protection goals, or standards. The development of waste-disposal standards is at a fairly advanced stage at the national level in several countries. Some provide rules that are explicitly formulated; others use basic principles in a more implicit form. Despite the diversity of the various national regulations, an international consensus on the basic principles for safety is now emerging. The following brief list of safety principles is extracted from a recent international consensus report on the subject.[16]

5.1 *Future Generations*

Two principles express the concern for future generations.

- The degree of protection afforded to future individuals shall not be less than what is acceptable today.
- The safety of a waste-disposal facility in the post-closure period shall not rely on the need for active steps (such as monitoring, surveillance, or remedial action) to be taken by a future society.

5.2 *Radiation Protection*

The overriding health concern related to radioactive wastes is potential harm from radiation. Although waste repositories are designed so that radioactive wastes will be isolated from the human environment for very long periods of time, it is recognized that radiation exposure is possible. Two main types of scenario by which this may occur have been identified. The first is by normal evolution, that is, due to the gradual degradation of barriers and the slow migration of radionuclides through the geological environment to the biosphere. The other type is termed a "disruptive scenario" and includes rare events, such as earthquakes, and also accidental intrusion into the site (e.g., mining). The principles covering radiation protection require that any risks to members of the public from these causes should not exceed those that would be acceptable on the basis of current international standards. These safety goals are quantitative targets which must be shown to be complied with in safety analyses of the waste repositories. (This is discussed later in Section 7.)

5.3 *Multi-Barrier Approach*

Of the technical criteria in the Safety Principles document,[17] only one is listed here because it is the overriding principle applicable to all kinds of waste disposal.

The long-term safety of a waste disposal facility shall be based on a multi-barrier concept and be assessed on the basis of the performance of the disposal system as a whole.

This principle calls for "defence in depth": there must be more than one barrier preventing or restricting waste from entering the human environment. Should one barrier fail, the others are still sufficient to ensure safety. A number of further standards addressing the various components of the system, such as waste form, design, site, etc., are then set down, but their implementation is always to be seen within the general framework of the overall systems approach.

5.4 *Compliance*

Finally, two principles address the means by which compliance with the safety principles can be shown.

- Compliance with the radiation protection principles shall be demonstrated by a safety assessment based on models that are validated as far as possible.
- A quality assurance program shall be established for all important parts of the waste disposal system.

The principles and criteria, briefly summarized in Sections 5.1 to 5.4, present a flexible framework and leave it to national authorities to fill in the details for certain components of the disposal system.

5.5 *Further Development*

Despite the advanced stage of the safety principles there is still room for further development. Three areas where additional work is needed may be specifically mentioned. First, regarding long-term aspects, safety assessment calls for predictive modelling of the future behavior of the disposal system; hence it encounters the difficulty of predicting into far future times. Some countries prefer to introduce a time cut-off of, say, 10,000 years; other countries are attempting to formulate specific long-term standards which should ensure that the radioactive releases, independent of future societal and environmental factors, are never greater than a small fraction of natural radioactivity releases.

Second, regarding human intrusion, it is difficult to assess the risk due to future humans intruding into a waste disposal facility, mainly because such an assessment involves prediction of a particularly unstable component of the system: human behavior.

Third, with respect to transboundary considerations, radioactive releases do not normally respect national borders; moreover, present borders may be meaningless in the long term. Hence harmonization of the national waste disposal standards and establishment of an international agreement on a minimum set of standards would be desirable.

6. Geological Disposal Concepts

It is convenient to discuss separately the disposal of the short-lived and long-lived wastes identified in Section 2, since generally different types of repositories are used for them.

6.1 *Mined Repositories for Long-Lived Wastes*[18]

The repository design is fundamentally intended to ensure adequate long term isolation of the waste and thus comprises a system of natural and engineered barriers to prevent or control the release of the waste into the biosphere. These barriers are designed to counteract or minimize the natural processes which could lead to release of the waste and, in some cases, also to minimize the effects of inadvertent human intrusion into the disposal zone.

In the majority of high-level waste repository designs the multi-barrier aim is achieved by a series of four components (Figure 1). The first is the waste form itself, which immobilizes the waste radionuclides in a solid matrix that is easily manageable prior to disposal and relatively stable under disposal conditions; thus it provides an immediate barrier which limits the rate of release. The waste will generally be sealed into a metallic canister and, in the case of vitrified high-level waste, will have been manufactured in such a container. This canister, perhaps enhanced by the addition of a second metallic or ceramic "overpack," will constitute the second of the barriers. When the containers are placed in the rock, it is necessary to fill the annulus of the emplacement hole with some form of backfill, both to ensure that the waste block is in thermal and mechanical continuity with the rock, and to minimize water movement around the container. This backfill can be designed to perform many physical and chemical functions to help contain the waste. It is generally called the buffer, and is the third of the barriers.

The final barrier, and the most important in many respects, is the host rock and geological environment in which the repository is constructed. This isolates the waste from the surface and controls water movement and geochemistry on both large and small scales. The first three barriers are within the engineer's control, and can be tailored to suit a particular concept or geological environment. Containers can be made as thick as required or designed to meet specific handling requirements. Waste forms can be selected at will, and buffers can be of any material or combination of materials. The waste matrix, container, and buffer are termed the "engineered barriers." The rock barrier is, of course, outside the engineer's control and can only be modified to a limited extent, such as by local rock grouting. The principal control on the rock barrier is exercised when the site is selected, and considerable geological research is therefore required to ensure an adequate understanding of the properties of rock types and environments considered potentially suitable for disposal purposes.

Many conceptual designs exist for waste repositories, but they are based on two main approaches. The simplest is where waste containers are placed in tunnels in the rock. Access to the tunnels is gained by vertical or inclined shafts. The tunnels are planned to be at depths of greater than 300 meters. The whole repository would be backfilled and sealed when the facility is decommissioned. In the second type of design, an extension of the first, canisters are placed in wide boreholes, drilled down from the tunnel floors, instead of being placed in the tunnels themselves. Figure 2 shows a typical high-level-waste repository design.

For a period of several hundreds of years after HLW has been buried, it will emit a steadily diminishing, but significant, quantity of heat. The host rock and engineered barriers must be capable of allowing this heat to diffuse away without losing their ability to act as barriers to radionuclide release and transport.

No facility for the disposal of high-level waste is yet in operation, although the WIPP facility in a salt formation in New Mexico is ready and awaiting a license to operate for the disposal of US military wastes containing long-lived radionuclides.[19] In Europe, plans are well advanced, and Sweden, Switzerland, and the Federal Republic of Germany plan to have operating repositories in the early years of the next century.[20] These time scales reflect, in part, the need to allow appropriate cooling of the vitrified waste and spent fuel before burial but also the time required for the extensive site investigation and testing programs required by national authorities prior to licensing.

6.2 *Repositories for Short-Lived Wastes*

Wastes of this type, also referred to as low- and intermediate-level wastes (with low concentrations of long-lived radionuclides) are currently being disposed of by emplacement in underground rock cavities and by means of shallow burial.[21,22]

Many of the considerations involved in the siting of these repositories are similar to those previously described for long-lived waste disposal. The following paragraphs describe existing or planned facilities in the Federal Republic of Germany, the German Democratic Republic, and Sweden.

In Sweden a disposal facility for low- and intermediate- level short-lived wastes has been built in bedrock. The facility SFR, the Swedish Final Repository for Radioactive Waste, opened in 1988 and is located close to the Forsmark nuclear power plant. The actual disposal chambers are located at about 60 meters depth under the sea-bottom about one kilometer off-shore (Figure 3). The waste, which is packaged before delivery to SFR, is disposed of in different kinds of rock-chambers. The choice of disposal chamber is based on the activity content of the waste and the handling equipment needed. The waste containing the highest activity is disposed of in the silo repository. This part of SFR consists of a cylindrical rock cavern in which a concrete silo with a height of 50 meters has been cast. The waste will be disposed of inside this silo and eventually surrounded by concrete grout. Between the concrete silo and the rock a one-meter-thick layer of bentonite clay is applied. Less active waste will be deposited in rock chambers. In one of the chambers the waste is deposited in concrete pits and surrounded by grout or covered with concrete. In the others, for very low-level waste, the waste is only emplaced with no extra backfilling.

Natural and engineered barriers thus isolate the waste from the environment. After closure of the repository it will gradually become filled with water by the natural inflow. The most important barrier is thereafter the host rock which has very low water transport. Situating the repository below the seabed is advantageous as the groundwater head is almost constant below the sea, which means that the groundwater flow rate will be very low. For the rock caverns the combination of practically stagnant water in the rock and the barrier functions of the waste packages and concrete structures provide adequate protection. In the silo repository the backfilling with bentonite clay acts as a seal to prevent groundwater from flowing through the silo. The only way that the activity can escape from the silo is thus by diffusion.

In the Federal Republic of Germany the present plans for disposal of low- and intermediate-level waste involve the use of the former iron-ore mine Konrad. In Konrad the waste will be disposed of in tunnels at more than 1,000 m depth. The choice of depth is based on the existence of a system of communication tunnels at this depth and a shaft. The actual disposal tunnels will, however, be excavated especially for this purpose. Most of the waste will be brought down in containers and stacked in the tunnels. After disposal the tunnels will be backfilled with the excavated material. The Konrad mine is very dry, water inflow is low, and the travel time to the environment for the groundwater is very long.

The following geological factors were of decisive importance in selecting the Konrad location as a site for a repository for radioactive waste that does not generate heat:

- The mine was well covered by a claystone-marlstone sequence several hundred meters thick and of low permeability.
- Because the ore horizon did not crop out anywhere in the vicinity of the Konrad mine, there was no direct contact between the deep groundwater and the groundwater near the surface.
- The barrier represented by the claystone-marlstone was not cut by any fault zones.

In the German Democratic Republic, a former salt mine is used for the disposal of low- and intermediate-level waste. Disposal started in the mine Bartensleben in 1979. Solid waste is deposited in the mine, although sometimes without packaging. Liquid waste is also disposed of directly into a cavern where it is solidified by adsorption and reaction with fly-ash. The main advantage of salt mines, for disposal, is the absence of water. It can be assumed that the main natural processes which can cause barrier deterioration and radionuclide transport are absent.

Burial of low-level radioactive wastes (LLW) in trenches or pits at shallow depths has been practiced for the last 40 years, principally by those countries in which nuclear programs were developed early.[23] The volumes of low activity wastes currently being generated are many orders of magnitude greater than high-level wastes, but because of their short-lived nature and relatively low potential hazard, a simple technique such as shallow burial has generally been considered an appropriate solution for their disposal.

In the early years of the nuclear industry, the shallow disposal practices were not well controlled and several examples of leakage of radionuclides from facilities in the USA have been reported.[24] These problems were mainly caused by inappropriate siting, in some cases leading to the trenches filling with water and overspilling, poor drainage, and inadequate covering. While none of the cases reported appears to have led to a radiological hazard, these practices are clearly unacceptable.

Nowadays, the wastes buried in shallow trenches tend to be subject to much more stringent controls in terms of their activities and radionuclide content, and in many countries this applies also to their physical and chemical form and packaging. Much more attention is being paid to siting and to engineered barriers and systems to minimize the chances of human intrusion and to prevent the leakage of radionuclides to the environment.

Shallow burial is practiced in many countries both for radioactive and non-radioactive wastes. Both the British and the French have examples of these facilities.

In the UK, low- and intermediate-level wastes have been disposed of at the Drigg disposal site since the 1950s. Drigg is located in the northwest of England about 500 meters from the sea. The waste is deposited in trenches based on glacial clay. The trenches are about 8 m deep x 25 m wide x 700 m long. Most of the waste is deposited without prior treatment. After deposition the waste is covered by at least 1 meter of soil. Up to now no drainage pipes have been included in the trenches. The clay bottom of the trenches, however, acts as a drainage system to protect the underlying regional groundwater. The trenches drain into a stream that runs off-site and discharges into the inter-tidal region of a nearby estuary. The water of the stream is monitored routinely for activity.

A program to improve the practices of disposal at Drigg is presently being implemented. The main changes will be as follows:

- The trenches will be engineered with concrete linings and underdrains.
- The waste will be compacted and containerized in metal boxes or drums before deposition. These will be stacked in a regular pattern in the trenches.
- Existing and future trenches will be capped with water-resistant materials (clay) to minimize the infiltration of water into the trenches.

The most elaborate shallow-ground disposal facility in the world is the French *Centre de la Manche* facility located close to La Hague on the Cherbourg Peninsula. The *Centre de la Manche* has been in operation since 1969 and has received some 300,000 m^3 of low- and intermediate-level short-lived waste. Two methods of disposal are used (Figure 4). Waste packages requiring a high degree of isolation are buried below ground and sealed in concrete to form monoliths. Those waste packages which do not require the same level of protection are then stacked on top of the monoliths to form tumuli. The stability of the tumuli are based on the use of very stable waste packages made of concrete. When disposal of the waste in a tumulus is complete, backfilling material (gravel) is poured over the entire stack to fill the gaps between the packages and to guarantee the stability of the tumulus. Once complete, the tumulus is then capped with a thick layer of impermeable clay.

The disposal structures at this facility perform a twofold role, to protect the packages against external interference, and in case of such an incident, to limit the consequences. This double function is provided by a series of barriers, the clay and earth of the tumuli, the gravel, the concrete in the monoliths, the packages themselves and the solidification matrix. The tumuli are made impervious to rainfall and underground water. A water collection and monitoring network situated at the base of the monoliths and tumuli enables the function of the facility to be checked. The extensive quality control program for the waste delivered to the Centre plays an important role in assuring the safety of the facility. At this facility, it is foreseen that control will be maintained for 300 years until the activity of the wastes has decayed to levels which would not give rise to unacceptable hazards. Beyond this time period the land can be returned to use as a non-active area.

7. Safety Assessment

In previous sections, the techniques being employed and planned for the disposal of radioactive wastes have been described. It is of course necessary to show that disposal facilities will perform as intended. This is not so easy, especially for long-lived wastes, since the time scales involved exceed any of the time frames normally considered in planning engineered structures. The safety assessment has to be based on predictive modelling of the long-term behavior of the waste-disposal system, supported by evidence obtained from laboratory and field studies and by analogy with the behavior of certain natural environmental systems.[25] However, it is recognized that because of the long time scales, proof of safety, in the strict sense of the words, is not attainable for waste-isolation systems.[26] The best that can be achieved is an analytical procedure that rigorously considers all of the elements of adequate isolation and, in that way, provides the basis for society to decide if the proposed waste disposal system is reasonable and sufficient and can be considered safe enough.

Considerable efforts have been expended in developing and proving models for waste-disposal systems.[27] In the following paragraphs the basic elements of a safety assessment are described, some typical results are presented, and the remaining problems are summarized.

The methods used for safety assessment are generally applicable to all types of underground repository. The following description of a safety-analysis methodology was developed for a shallow-ground facility, but the principles remain fairly generally applicable. The disposal system can be described as a combination of the following components:

- waste type, waste form and, where applicable, stabilizer, container, overpack, and/or migration retardant;
- repository and its engineered barriers, including backfill of excavations and ground cover;
- near-surface geosphere (sediments, rocks) at the repository site; and
- human environment (soil, surface waters, shallow aquifers, atmosphere, biota).

The safety analysis of such a system has the following three basic components:

- identification of phenomena that could lead to a release of radionuclides, influence the rates at which releases occur, or influence the rate of transport of radionuclides through the environment;
- estimation of the probabilities of occurrence of these phenomena and quantification of their effects on the disposal system;
- calculation of the radiological consequences of releases (that is, doses to individuals and populations and, if required, estimation of subsequent health effects).

The first two items usually are considered under the heading of "scenario analysis" and the third item usually is referred to as "consequence analysis."

7.1 *Scenario Analysis*

The phenomena that are potentially relevant to scenario analysis for repositories in shallow ground are human activities (for example, construction, farming, drilling for mineral resources); natural processes and events (erosion, flooding, groundwater movement); and waste-repository processes (gas generation, mechanical disturbance of soil or rocks at the repository site).

The results of scenario analyses consist of the definition of scenarios (that is, the release and transport parameters required for consequence analyses) and the estimates of the probabilities of occurrence of these scenarios as a function of time. When scenario analyses are carried out for specific disposal systems and sites, it will usually be possible to eliminate some potentially relevant phenomena from detailed consideration, either because they have very low probabilities of occurrence or because their effects can be shown to be insignificant.

7.2 *Consequence Analysis*

After the repository release scenarios have been defined, the consequences to humans have to be estimated. Calculation of consequences requires the development of a computational system that enables the transport of radionuclides through the environment to humans to be modelled and radiation doses to be assessed.

As a first step, a prediction of radionuclide release rates from the repository is made, followed by estimates of the radionuclide concentrations in the various compartments of the environment. The second step consists of a prediction of transport rates of released radionuclides among various compartments and humans. The third step involves a prediction of radionuclide interactions with humans for each scenario identified during scenario analysis.

Consequence analysis involves following the progress of the released radionuclides to humans via different pathways--for example, through drinking water and the consumption of aquatic foods, bathing and swimming, and more indirectly through plants and domestic animals. (Plants can accrue radionuclides via irrigation or through topsoil.) Results of the overall safety analysis may be expressed in different ways depending upon the radiological criteria to be satisfied, for example, in terms of probabilities, individual doses, individual risks, and collective doses and risks.

7.3 *Reliability of Model Predictions*

Since the safety assessment of radioactive waste repositories is heavily dependent on the use of mathematical models, it is necessary to consider how much confidence can be put in their predictions. All environmental models necessarily have uncertainty associated with their predictions since they are simplified representations of complex environmental systems. The problem of model uncertainty is particularly difficult to resolve in relation to repository safety assessments. This is because directly relevant experimental results and environmental measurements for use in validation are hard to obtain, especially in relation to the long periods over which the models are applied. There are, nevertheless, a variety of approaches that can be applied by the safety assessor to increase confidence in results of model application.

- Ideally, a model should be validated by means of comparisons of its predictions with appropriate observations. Full validation of all models used in waste repository safety analysis is not possible, but partial validation of submodels can be achieved by the use of laboratory and field experimental data.
- In some cases, less controlled types of validation can be performed by making comparisons with natural systems that contain naturally occurring radionuclides.[28,29,30] These have the advantage of providing real evidence of transport processes over the very long time scales that must be considered in relation to underground waste disposal. Difficulties with this type of validation are that experimental conditions are not generally well understood and may not be directly relevant to the environment surrounding the repository. In the case of the Cigar Lake uranium deposits in Canada there is a close analogy with an actual human-made repository, and one can make comparisons between the two cases.
- One method of circumventing the problem of model uncertainty is by the use of "worst case" models. This approach gives no indication of the reliability of prediction, but provides confidence that there is little chance of underprediction. It may be used to demonstrate that doses are well below limits as, for example, in the generic studies referred to earlier.
- Model comparison exercises have a valuable role in safety assessment. They enable the numerical accuracy of modelling codes to be checked, but in

addition they may be used to compare different modelling approaches for predicting transfer in defined circumstances. This often results in valuable insights being obtained. Various international model comparison exercises relevant to underground waste disposal are taking place at the present time; for example, the Swedish sponsored INTRACOIN (geosphere far-field radionuclide transport models), HYDROCOIN (groundwater transport models), and BIOMOVS (biospheric radionuclide transport models).

- Statistical techniques may be used to evaluate what effect uncertainty about the values of individual model input parameters has upon the uncertainty of the overall model predictions (Figure 5). It has to be borne in mind, however, that this type of uncertainty analysis does not provide a check on the validity of the model itself.

Demonstrating the reliability of predictions of radionuclide transfer from underground repositories is not easy and is often an indirect process. Confidence in model predictions has to be built up by the combined use of a variety of techniques, such as validation, model comparison, and parameter uncertainty analysis.

In summary, the results of safety analysis indicate that the radiological impact of waste repositories will be small and that compliance with radiological protection standards can be achieved (Figure 5).[31,32] It is nevertheless recognized that there are considerable uncertainties associated with these assessments, mainly because of the long time scales involved. Many of the current research efforts in this field are aimed at reducing uncertainties.

8. Disposal of Other Types of Hazardous Waste

The list of substances which are now considered chemically or biologically harmful is increasing as our understanding of the medical effects of chemical pollutants grow. The quantities now classed as hazardous are large compared with the arisings of radioactive wastes; in the UK the annual output of hazardous wastes are around 50 million tonnes while the quantity of low-level radioactive wastes is around 100 thousand tonnes per year.[33]

It is clear that very different standards are being applied to radioactive and non-radioactive hazardous wastes in the UK. Similar evidence is available from other countries. The reasons are partly historical, but the persistence of the anomaly is less easy to understand. There are now indications that the balance is slowly beginning to be redressed, by making more stringent demands for the disposal of chemical wastes. The lower standards which have been applied to non-radioactive wastes in no way justify a relaxation in standards for the disposal of radioactive wastes but merely provide a perspective within which current national plans for radioactive waste disposal may be seen.

9. Summary

More than 99% by volume of the radioactive wastes arising from the nuclear fuel cycle come into the categories for which there are well established disposal techniques. For low-level, short-lived wastes the preferred disposal method is shallow burial, and numerous operating facilities of this type exist. A distinction may be drawn here between carefully controlled sites of the modern type, which have been licensed only after rigorous site selection and analysis procedures and in which segregated wastes are packaged and protected against inadvertent intrusion, and the practices of the fifties and sixties in which little care was given to the waste-disposal operation. For intermediate-level, short-lived wastes the currently preferred options are shallow burial in engineered structures or emplacement in near-surface rock cavities. Wastes of this type are being routinely disposed of in some countries, but in others they are being stored at nuclear power plants awaiting disposal, usually because of problems in gaining public acceptance of the disposal sites.

For the remaining wastes, the high-level, long-lived types, no operating sites yet exist, but there is a general international consensus that deep geological disposal is the most appropriate method, and that isolation can be best achieved by means of a multi-barrier approach in which there are several independent barriers, both engineered and natural, to prevent the radionuclides from the waste reaching the biosphere. The first repositories of this type are planned to be operational in the early years of the next century. Currently these types of waste are being stored, either as spent fuel rods, as liquid products from reprocessing, or as glass blocks containing the reprocessed material. A cooling period of a decade or more is desirable for these wastes before final disposal.

The many safety assessments which have been carried out to date on deep geological repositories indicate that the associated risk to persons and their environment will be very small indeed, a reflection of the considerable safety margins being designed and engineered into these repositories. Although an absolute proof of safety at very long time scales is not achievable, the evidence presented in this type of assessment should provide a sufficient degree of understanding of the disposal system and its long term behavior to permit national authorities to make judgements on the adequacy of the safety of proposed disposal facilities.

The resistance of local communities to having radioactive waste disposal sites located near to them is understandable, given the fears of radiation generated by past events, but unjustified in the light of the low level of hazard that the most pessimistic safety analysis would reveal. Another important perspective on the waste-disposal issue is obtained through comparisons with current practices in hazardous waste disposal. These comparisons show that the standards being applied in the disposal of hazardous wastes, containing toxic chemicals and heavy metals, are often very much lower than those for radioactive wastes, even though the potential hazards are at least as great as those associated with the majority of radioactive wastes.

Table 1

RELATIVE QUANTITIES OF THE CATEGORIES OF RADIOACTIVE WASTES AND OF OTHER HAZARDOUS MATERIALS IN THE UK

Waste type	Arisings (M Tonnes)	
Radioactive waste:		
low level	1.0	
intermediate	0.16	total to year 2000
high level	0.004	
Industrial wastes containing hazardous materials	5-10	per year
Spoil from coal mining	50	per year
Refuse (England and Wales)	20	per year

FIGURE I: SCHEMATIC ILLUSTRATION OF A MULTI-BARRIER SYSTEM OF CONTAINMENT FOR HIGH-LEVEL WASTE

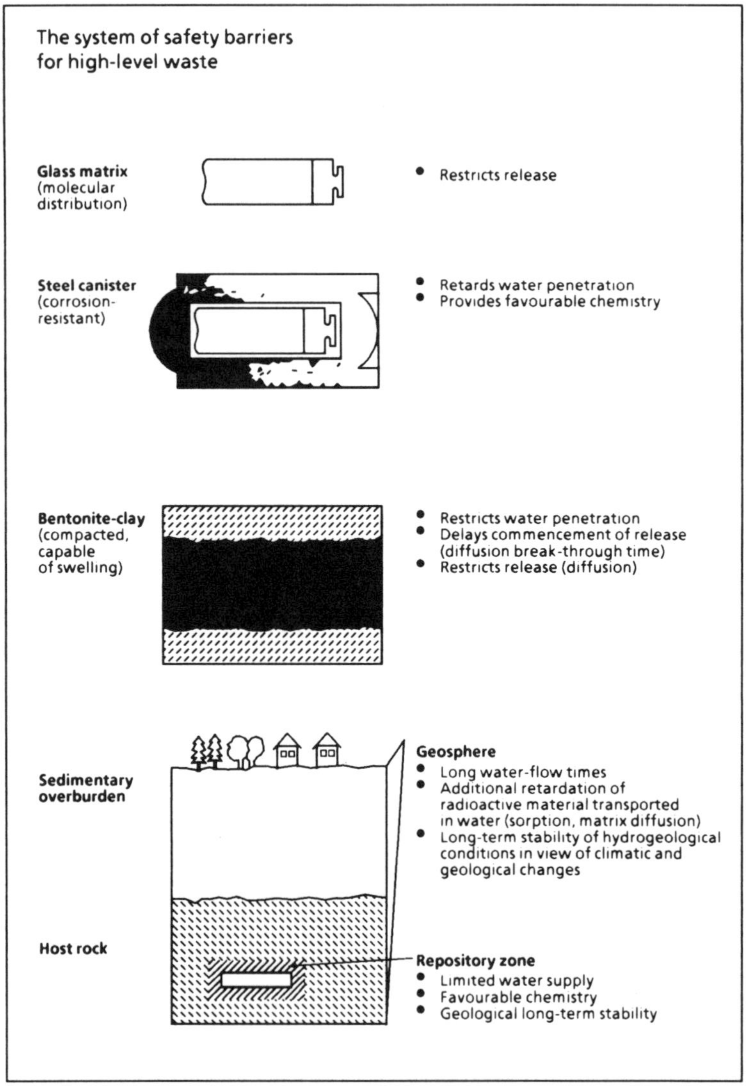

FIGURE II. CONCEPTUAL DIAGRAM OF A MULTI-LEVEL HLW REPOSITORY IN HARD BASEMENT ROCKS

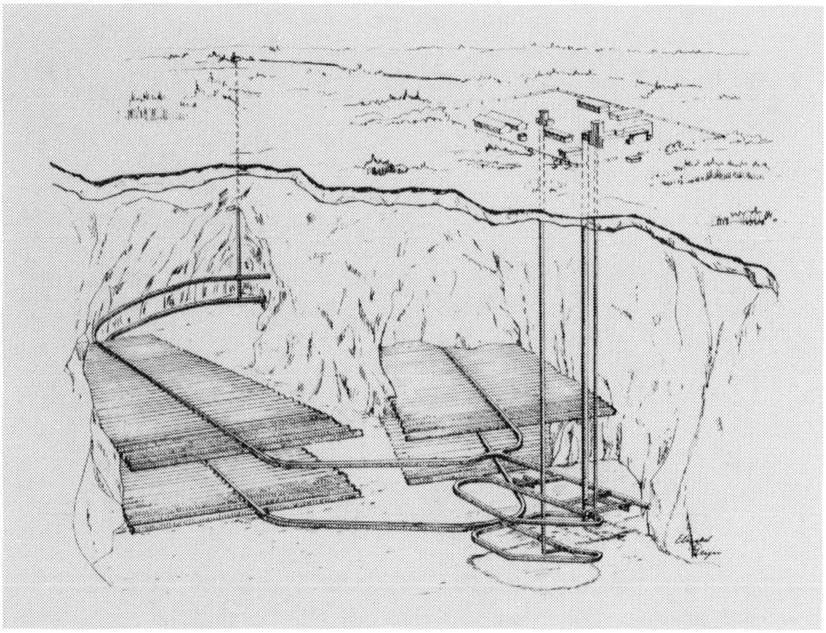

Courtesy of SKB, Sweden
Drawing by Elisabet Geijer

Figure III: The Swedish Final Repository for Reactor Waste. The tunnels are 160 m long and the cylindrical cavern 70 m high

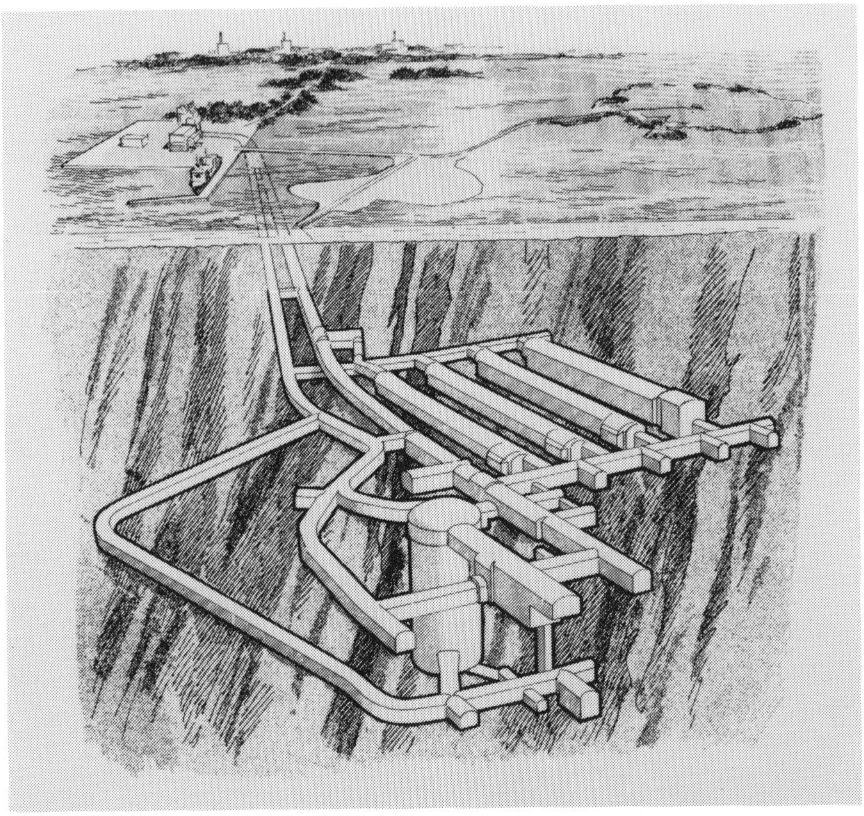

Courtesy SKB, Sweden
Drawing by Elisabet Geijer

FIGURE IV: ILLUSTRATION OF EARTH-MOUNDED DISPOSAL FOR SHORT-LIVED WASTES

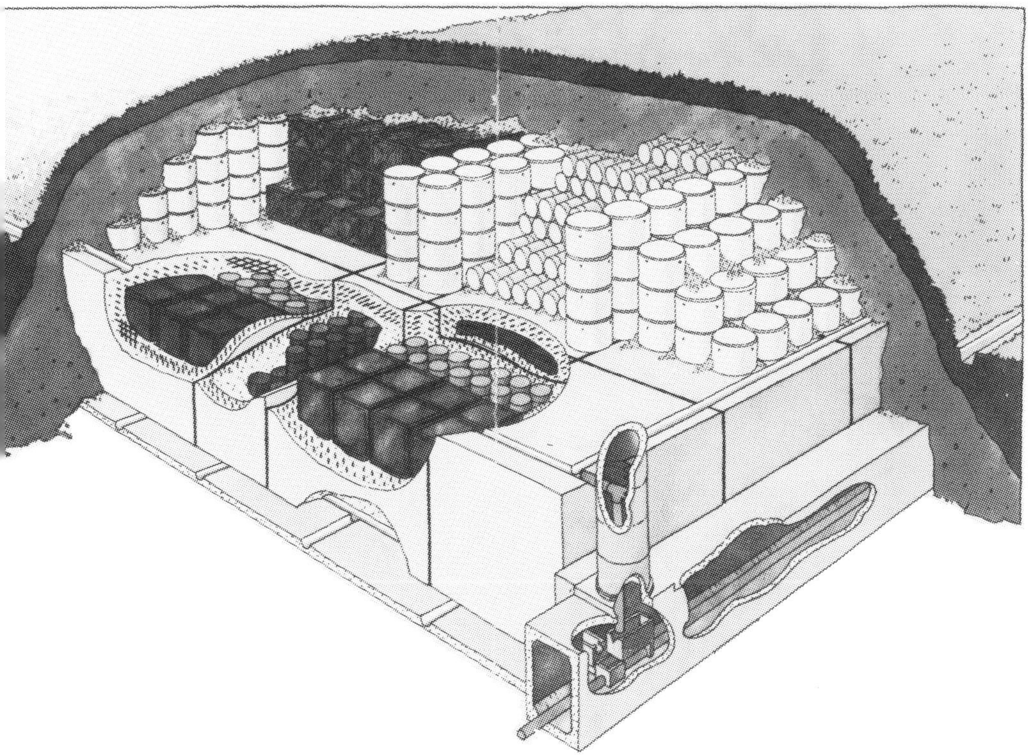

Courtesy of ANDRA, France

FIGURE 5A: SUMMARY OF RESULTS FROM UNCERTAINTY ANALYSES. NORMAL EVOLUTION OF REFERENCE SITES

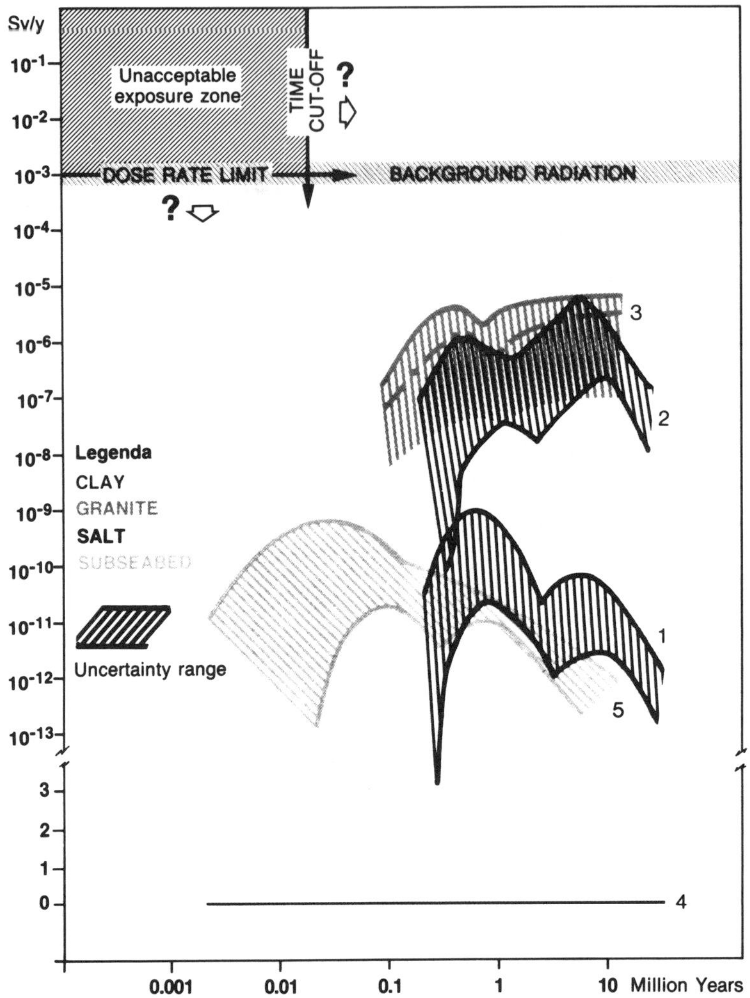

(from PAGIS report, Note 3, courtesy CEC)

Note: the authors of this study point out that while comparison of the results with natural background radiation is legitimate, comparisons between the options themselves may not be appropriate because of difference assumptions made.

FIGURE 5B: SCENARIOS CONSIDERED

Clay
1 River pathway. Curves are 50th and 95th percentile of dose probability density function (pdf) 2 Well pathway. Curves are 50th and 95th percentile of dose pdf.
Granite
3 Curves are **mean dose** (lower) and 95% confidence bound based on Tchebycheff's inequality (see Chapter B, Methodology)*
Salt
4 No release in the normal evolution scenario.
Sub-seabed
5 Curves are 50th and 95th percentile of dose pdf

NOTES

Note: This review chapter draws heavily upon the book cited in note 11.

1 World Commission on Environment and Development, *Our Common Future*, Oxford, Oxford University Press, 1987.

2 Nuclear Energy Agency (NEA) of the Organization for Economic Cooperation and Development (OECD), *Technical Appraisal of the Current Situation in the Field of Radioactive Waste Management*, Paris, OECD, 1985; hereafter cited as: NEA 1985.

3 Commission of the European Communities (CEC), *Performance Assessment of Geological Isolation Systems*, *PAGIS*, Summary Report, EUR-11775, Brussels, CEC Office of Official Publications, 1988; hereafter cited as: CEC 1988.

4 Nationale Genossenschaft fur die Lagerung radioactiver Abfalle (Nagra): *Projekt Gewahr 1985*, 8 Projektberichte NGB 85-01.....08, Baden, Switzer land, National Cooperative for the Storage of Radioactive Waste (NAGRA), January 1985. (Condensed English version as Project Gewahr, project report NGB-85-09, June 1985.)

5 KBS-3, *Final Storage of Spent Nuclear Fuel*, Stockholm, Sweden, Swedish Nuclear Fuel Supply Company, SKBF, May 1983.

6 N. A. Chapman and G. M. Williams, "Hazardous and Radioactive Waste Management: A Case of Dual Standards," in J. R. Gronow, A. N. Schofield, and R. K. Jain (eds.), *Land Disposal of Hazardous Waste*, Chichester, U.K., Ellis Horwood, 1988, p. 259; hereafter cited as: Chapman and Williams, 1988.

7 Swedish National Institute of Radiation Protection (SNIRP), *Disposal of Radioactive and Other Hazardous Wastes*, Proc. International Workshop, Stockholm, June 1988, Ds 1989:20, Stockholm, Ministry of Energy and Environment, January 1989; hereafter cited as: SNIRP 1989.

[8] UK Radioactive Waste Management Advisory Committee, *5th Annual Report*, London, Her Majesty's Stationary Office, June 1984.
[9] SNIRP 1989.
[10] UK National Radiological Protection Board, *Living With Radiation*, London, Her Majesty's Stationary Office, 1986.
[11] N. A. Chapman and T. G. McKinlay, *The Geological Disposal of Nuclear Waste*, John Wiley and Sons, 1987; hereafter cited as: Chapman and McKinlay, 1987.
[12] NEA 1985. [13] Chapman and McKinlay, 1987.
[14] M. S. Bedinger, *The Hydrologic Aspects of Land Disposal of Radioactive Waste*, UNESCO, Technical Documents in Hydrology, Paris, UNESCO, 1987.
[15] Nuclear Energy Agency of the Organization for Economic Cooperation and Development, *Review of the Continued Suitability of the Dumping Site for Radioactive Waste in the North-East Atlantic*, Paris, OECD, 1985.
[16] International Atomic Energy Agency (IAEA), *Safety Principles and Technical Criteria for the Underground Disposal of High-Level Radioactive Wastes*, Safety Series No. 99, Vienna, IAEA, 1989; hereafter cited as: IAEA 1989.
[17] IAEA 1989.
[18] Chapman and McKinlay, 1987.
[19] J. B. Tillman, "Meeting the Challenge of Radioactive Waste Disposal at the WIPP," *Waste Management* 89, Roy G. Post (ed.), Proc. Symp. at Tuscon, Arizona, February 1989, Washington, D.C., Library of Congress No. 86-644175, 1989.
[20] International Atomic Energy Agency, *IAEA Bulletin* 31, No. 4 (1989).
[21] International Atomic Energy Agency, *Management of Low and Intermediate Level Radioactive Wastes*, Proc. Symp. IAEA/CEC, Stockholm, May 1988, STI/PUB/784, Vienna, IAEA, 1989.
[22] International Atomic Energy Agency, *Siting, Design and Construction of Underground Repositories for Radioactive Waste*, Proc. Symp., Hannover, March 1986, Vienna, IAEA, 1986.
[23] Chapman and McKinlay, 1987.
[24] International Atomic Energy Agency, *Shallow Land Burial of Low-level Radioactive Wastes in the USA*, IAEA-SM-243/152, Vienna, IAEA, 1980.
[25] CEC/IAEA/OECD/NEA, *Safety Assessment of Radioactive Waste Repositories*, Proc. Symp., Paris, October 1989, to be published by OECD, Paris; hereafter cited as: CEC, *et al.*, 1989.
[26] CEC, *et al.*, 1989. [27] CEC, *et al.*, 1989.
[28] Chapman and McKinlay, 1987. [29] CEC, *et al.*, 1989.
[30] Swedish Nuclear Power Inspectorate, *Geoval*, *87*, Proc. Symposium on Verification and Validation of Geosphere Performance Assessment Models, Stockholm, April, 1987, Stockholm, Swedish Nuclear Power Inspectorate, 1988.
[31] CEC 1988.
[32] CEC, *et al.*, 1989.
[33] Chapman and Williams, 1988.

Nuclear Technology Today: Promises and Menaces

Bena-Silu
Kimbanguist Church
Kinshasa, Zaire

1. Introduction

Nuclear technology is, without question, a science and technology of the twentieth century, although it was in 1886 that Henri Becquerel discovered radioactivity, a nuclear phenomenon. Nuclear technology was developed around the middle of this century and has become one of the most marked scientific and technological characteristics of our era.

Less than a hundred years ago, Albert Einstein formulated the concept of (relativity) mass-energy. Nuclear science has since confirmed it. The growth of nuclear reactors today is testimony to the remarkable and spectacular progress of nuclear power.

2. Progress in Controlled Nuclear Fusion

In examining history, it appears that in many cases, several centuries have occurred between the discovery of a scientific concept and its transformation into a technological reality. Progress in controlled nuclear fusion is so fast, however, that it undoubtedly will short-cut the predicted estimate for putting into place the first commercially functioning thermonuclear fusion reactor around the years 2030 or 2050. Ever since 1986, the Joint European Torrus, the strongest of the existing tokomaks in the world today, has been able to produce (for extremely brief periods of time) an amazing (internal) temperature of 140 million degrees Celsius, which is more than nine times that which exists at the center of the sun. Since the Tokomak Fusion Test Reactor (at Princeton University in the US) came into service in December 1982, and especially with the JET (June 1983), the realized progress in thermonuclear controlled fusion, according to the magnetic confinement approach, has been more rapid in four years (1982-86) than in the 39 years since 1952.

The best value of fusion products attained in the JET and TFTR has improved by a factor of 1000, just in the space of 20 years. The previewed stages of the next European Torus du DEMO and of ITER, without doubt, will attain more than technical feasibility in this exciting field of nuclear research, despite the incredible technological obstacles which must be overcome. The challenge is reminiscent of the goal, faced by twentieth-century scientists, of walking on the moon. Nuclear technology has now attained the age of reason, but not without triumphing over the many snags that have marked out its path since the time when Enrico Fermi constructed the first atomic pile during the Second World War.

3. Accidents and Nuclear Safety

Of the many known nuclear accidents, there are three that bear mentioning: Windscale (United Kingdom, October 1957), Three Mile Island (US, March 28, 1979) and, above all, Chernobyl (USSR, April 27, 1986). These spectacular accidents confirmed that no technology is infallible, no matter how advanced it is. Smaller accidents happen almost daily in many nuclear reactors all over the world. The fact that major nuclear-reactor accidents are rare cannot justify the lessening of measures for reinforcing nuclear security. The consequences of nuclear accidents pose such a great danger to all life and to the environment that they must always be anticipated as much as possible. Unfortunately, no one can predict exactly when an accident might happen.

Moreover, it is well recognized that accidents can be caused by faulty technology or by human error, if not both at the same time. Many nuclear-safety studies have been done especially in the countries which have had a long nuclear-technology experience and by first-class organizations, such as the International Atomic Energy Agency.

Chernobyl demonstrates, however, that there are exceptional, or even irreversible, risks associated with nuclear power. At the technical level, nuclear-safety studies have brought about considerable improvements in the emergency core cooling systems (EMCCS), systems which rapidly detect leaks of coolants, neutrons, actinides, or other fission products. Emphasis is also placed today on the quality of parts and accessories which enable a reactor to function with as few failures as possible. Thus, we have produced a nuclear-fuel rod more resistant to heat-caused deformations, an improvement which results in better energy efficiency.

As far as fast neutron reactors go, fast breeders (of which the best prototype is the 1200-Mwe Superphenix in France) still have many unknown safety factors to be resolved. Insofar as this reactor is the most powerful of commercial fast breeders operating in the world, nuclear physicists and engineers worldwide follow its successes and failures with great interest. The fact that fast breeders produce more fissile material than they consume should have motivated many countries to build a larger number. However, that is not yet the case for a good number of reasons: high construction costs; a technology which should still be observed for several years; and hostile public opinion vis a vis nuclear production in general.

It goes without saying that a serious chemical explosion in a large fast breeder like the Superphenix could have extremely serious consequences for the surrounding population and for the environment. Such an accident, if it is not controlled in time, could lead to a meltdown and release of actinides into the atmosphere. Difficult, but not insurmountable, technical problems have also restrained the spread of fast-breeder reactors. In fact, this reactor needs 400 more neutrons than conventional reactors in order to produce a fission reaction. In a fast breeder, one has to minimize the loss of fast neutrons before they cause chain reactions. Therefore, the core of a fast breeder has to be more compact than those of other reactor types; as a result it does not have a moderator.

Another constraint is the liquid-metal (usually sodium) coolant. Sodium is flammable in contact with air. This makes it extremely difficult and hazardous to put out a major fire in a sodium-cooled reactor. Since it boils at 99 degrees Celsius, sodium does not need to be under pressure in the fast breeder; this solves at least one complex engineering problem. On the other hand, the opacity of sodium makes it especially difficult to control visually its circulation in the cooling circuits. When the reactor is shut down, sodium should never be allowed to cool below 97 degrees Celsius (its fusion temperature), to avoid solidification. When sodium (Na23) absorbs neutrons, it becomes Na24, a fierce gamma emitter. Finally, sodium reacts explosively with water, making severe precautions necessary.

Generally, fast breeders use liquid-sodium coolant in two circuits: one primary circuit which is completely inside the core, and a secondary circuit which transfers heat to heat exchangers. The energy produced by a fast breeder is around 41.4%. One can see that this type of reactor carries high risks. To date, one Superphenix has done well, and there is discussion of building a Superphenix II of 1500 MW.

The Three-Mile-Island and Chernobyl accidents have brought to light the need to take the human factors more seriously into consideration in questions of nuclear safety. In effect, the operators can make a wrong judgment, due to insufficient technical education. They can provoke or aggravate a nuclear accident. When and which decisions may an operator make when faced with an abnormal development in a nuclear reactor?

Reactors of the current generation are of better quality than their predecessors. In the light of accumulated nuclear experience, recent reactors are better conceived than earlier ones. The current tendency is to simplify the design of reactors, be they conventional or fast breeder, cooled with heavy or light water, boiling-water reactors or pressurized-water reactors. All these innovations require engineers and technicians to take regular refresher courses or become obsolete. Besides universities, a good number of nuclear contractors and operator firms have set up their own educational establishments to provide their personnel with continued education. *Electricité de France* (EDF) has done this. In addition, the variable safety requirements of a reactor have lead to the installation of new and more efficient mechanisms for making urgent decisions.

One major shortcoming of nuclear safety is the diversity of reactors, i.e., the fact that there are too many different types of reactors although, in every case, control rods must be quickly inserted in the reactor core to stop fission reactions.

For each type of reactor, there is a different safety system. For large companies which simultaneously handle different reactor types, this translates into high costs in education and exploitation.

We should immediately reject assertions that one reactor type is safer than another since, if an accident occurs in a reactor, at least three factors are unforeseeable: when and where the reactor accident will occur, the time of the accident, and when the operator will interact with the reactor. It is a fact, however, that light-water reactors are dominant in the world. This implies neither that this type of reactor is better, nor that it is safer than its competitors.

4. The Management of Nuclear Waste

Because nuclear-fission products have a very long radioactive half-life, they constitute one of the most controversial aspects of the nuclear question. The issue of waste, however, cannot be treated outside the general problem of the nuclear fuel cycle, from the mining of radioactive minerals (uranium and thorium), milling it, enriching it, making it into fuel, introducing it into the reactor, treating it in a specialized plant to separating the wastes, and storing them. During the complicated process which involves transportation, especially strict precautions must be taken against radiation. Certain parts of the process require remote-control technology.

In the 1960s, nuclear power was regarded as "clean energy." Today, public opinion has come to believe this is untrue: nuclear waste pollutes the environment for a period of time longer than several human generations. A number of countries have either slowed or given up their nuclear-power construction program, because of the problem of nuclear waste. We should note in passing that the treatment of a ton of uranium minerals produces 3,700 liters of toxic and radioactive liquid. About 250,000 tons of uranium minerals have to be treated to produce 1000 tons of uranium. The risks of radioactive contamination start from the mining of uranium until the storage of nuclear waste. According to some sources, the US nuclear military program has led to the accumulation of vast amounts of mineral waste, perhaps amounting to 100 million tons. However, the other nuclear countries do not come out any better in this matter. The disposal of nuclear waste remains a real question, even for countries which possess powerful research reactors.

Besides the safety issue, there is the high cost incurred in the storage of the radioactive wastes, certain types of which must be kept for several months in underground pools until their radioactivity has decreased sufficiently to permit their transfer to other locations. To the problem of the costs caused by the storage of radioactive wastes is added that of their recycling, a task so delicate that it must be undertaken by remote control because of the toxicity and strong radioactivity of the wastes.

Nuclear wastes are separated into high-level and low-level radiation wastes. In the "final" stage, some of them are stored in deep geological locations, if they are not simply deposited in the ocean depths, despite an international treaty forbidding such practices. The effective action of ecologists in numerous states has forced the nuclear industry, and even governments, not to consider dumping nuclear wastes in the seas.

In the Federal Republic of Germany, the wastes are stored in abandoned salt mines. In France, remarkable advances have been made in storing solid nuclear wastes by transforming them into borosilicate. However, it seems that this solution is not yet standardized throughout the world.

5. Proliferation and Transfer of Nuclear Technology

On proliferation, it should be noted that, several decades ago, the advanced countries of the northern hemisphere lost their monopoly of the knowledge of nuclear science and technology. In our day, there are fewer and fewer universities in the world that have not set up a chair of nuclear physics. Although at the end of World War II, the US had the sole mastery of nuclear science (with the aid of a number of scientists who came from elsewhere), now in the majority of countries, there are nuclear experts. The list of countries with a mastery of nuclear science grows every year, even to the extent of making and exporting entire nuclear reactors.

A number of interesting proposals have been made by many experts, including Taylor and Feiveson, to prevent the diversion of recycled fuel to military ends. These two American researchers proposed that the reactor types be limited to two: one (the principal one) made up of thermal reactors working with thorium, and the second (the less important) made up of fast breeders which partly will use thorium in their cores, so as to produce U-233 for thermal reactors. In the thorium thermal reactors the ratio of U-233 to U-238 should be 1:6, or such that, even after the chemical separation of thorium, plutonium, and fission products, the uranium obtained is not of a quality able to be converted for military ends. From the practical point of view, these suggestions have no chance of being upheld because no authority exists that is able to enforce them. This is all the more true as more and more countries are resorting to nuclear technologies, based upon national expertise. Besides, the chemistry for separating thorium, uranium, and plutonium by remote control presents an extremely difficult challenge.

Much more serious is the rapid conversion of civilian nuclear technology to that for fabricating nuclear weapons. This is the path which has been followed by four countries (including South Africa) for making nuclear weapons. It is reported that soon Brazil, Argentina, South Korea, and Taiwan will have accumulated sufficient knowledge to produce nuclear weapons. That list, of course, is not comprehensive. It is not always possible to stop nuclear proliferation. The very unfair Nuclear Non-Proliferation Treaty (NPT), drawn up by the large nuclear powers under the aegis of the International Atomic Energy Agency, cannot prevent a determined country from developing nuclear weapons - all the Treaty can do is slow down the process of acquisition. Of all the NPT initiating nations, none has committed itself not to resort to nuclear weapons against any nation which does not possess nuclear arms. During the Falkland-Islands war in 1982, Britain brought nuclear weapons into the war area that could have been used against Argentina, a non-nuclear power.

Moreover, there are discriminating provisions in the NPT that protect the nuclear hegemony of power, of nations already possessing a nuclear arsenal. One can understand why nuclear experts in developing countries express our total dissatisfaction with the NPT. In numerous meetings, experts and government representatives from major nuclear countries express annoying paternalistic opinions to our young nations, nations that do not possess adequate energy resources for economic development. These countries say that the creation of nuclear industries is associated with very serious problems, notably involving nuclear waste reprocessing. Such proliferation problems are real, and the nuclear experts of developing countries neither ignore nor underestimate them. Yet, despite all their complexities, it is up to our nations to live with the consequences of our own decisions. In general, if we oppose proliferation of nuclear weapons, it is not for the same reason as those in developed countries, who act on the basis that they must retain control over certain technologies. A nation determined to produce nuclear weapons can always have them, in defiance of the NPT provisions.

The decision made by President Carter, to suspend reprocessing of nuclear waste, is in line with the nuclear-powers' attempt to protect their hegemony in nuclear armaments. Such a moratorium did not meet its objective since, as mentioned previously, four nations, including South Africa (in the last case, with western assistance) have become, beyond all doubt, military nuclear powers. A mass of 14 kg of U-235, enriched to more than 99%, will suffice to produce an atomic bomb. For Pu-239, 6 kg are what is needed for a bomb. Several countries have a civilian nuclear industry which produces sufficient fissile material to make many atomic bombs.

Because of the extraordinary costs, the vast size of required facilities, and the large number of needed experts, the gas-diffusion enrichment technique for U-235 has little attraction for countries seeking to acquire a nuclear-weapons capability. Centrifuge technology, although a difficult technology, is the shortcut to uranium. With some imagination, it is possible to obtain (on the open market) equipment needed to enrich uranium, utilizing the centrifuge technique. Enrichment using the laser technique has been established in many laboratories in technologically advanced countries. To this day, the results do not seem to have justified the efforts. It nevertheless could become otherwise tomorrow.

6. Radiological Protection and Notification of Serious Accidents

In general, most countries with nuclear installations abide by radiation-protection standards laid down by the International Commission on Radiological Protection (ICRP). There are countries who apply even tougher standards. Following numerous incidents everywhere, nuclear installations have been closed temporarily or permanently by bodies responsible for non-observance of safety standards.

The accident at Three Mile Island was nearly as bad as the one at Chernobyl. In both cases, the reactor core melted, and nearby populations were threatened by fallout emanating from the crippled reactor. In August 1989, three years after the Chernobyl accident, it was reported that, at one time, Soviet authorities had envisaged, and then rejected, evacuating a population living in a village 400 km from Chernobyl.

7. Decommissioning Nuclear Installations

Large reactors, having reached the end of their working lifespan, must be decommissioned as part of the cost paid by all countries embarking on a nuclear program. Electricity-generating or research reactors, fuel-fabrication, or reprocessing plants are the installations destined to need decommissioning. There are also cases of grave accidents (Three Mile Island and Chernobyl) which led to premature decommissioning of a nuclear installation. Generally speaking, the technology for so doing is reliable, but is limited to countries with a lot of nuclear experience.

Statistics from the International Atomic Energy Agency (IAEA) indicate that between 1985 and the year 2010, "reactors equivalent to more than 50 electric gigawatts (GWe) could be ready for dismantling, only in member countries of the OECD. The extraordinary spring of nuclear industry in the world, in particular in the northern hemisphere, shows that the number of nuclear installations to be decommissioned will grow continuously. Not all nuclear countries possess the same degree of expertise as far as the dismantling of these installations are concerned. For this reason, the IAEA, as well as other countries, in virtue of bilateral or multilateral agreements, are bringing them assistance. Until 1985, some 135 installations in 16 countries were being decommissioned. The dismantling of every nuclear installation should be made with a particular methodology, one whose main steps include: calculation of residual radioactivity in the installation; decontamination and dismantling; and the sorting of wastes in view of their definite disposal or reutilization.

It is important to note that, during the whole period of dismantling, one must keep monitoring the radioactive doses continually, to assure that no danger is posed to the safety of the population. The dismantling of a reactor has to be made in stages, and according to the norms of radiological protection. There is no doubt that the experience acquired by the nuclear engineers during the operation of decontamination and dismantling of the TMI-2 reactor is unique in the world. The reports on this subject communicated to the IAEA are therefore extremely useful to the global nuclear community. To underline the obtained success, it should be mentioned that, at Three Mile Island they have succeeded in removing the cover of the damaged reactor and taking the reactor core out of the biological containment. During these complex operations, the safety norms have been met, even to the point that the groups of experts, protected by all the radiological equipment, were able to go in and out of the building.

For decontamination of the damaged reactor at TMI, there had to be developed original techniques which, until then, never had been applied. Organic resins were used to absorb the fission products; this allowed the decontamination of more than 20,000 m^3 of water. By 1980, 43,000 Bq. of krypton had been treated by the method called "SDS" (submerged demineralization system) as well as 22,713 m^3 of highly contaminated water. A better knowledge of ionizing radiation and the way to treat it, in the context of a major power-reactor accident, has thus been acquired. Decontamination and dismantling operations that take place at the TMI-2 reactor are the most complex ones ever used in the world. Informatics and robotics, as well as TV techniques, have contributed to their success.

Today, as we look retrospectively at the core melts at Three Mile Island and Chernobyl, we could say that they have increased our nuclear knowledge of engineering, decontamination, and dismantling. For Chernobyl, the USSR has not announced any dismantling plan, although the eventual plan should allow us a better knowledge about how to deal with ionizing radiation in the context of a major power-reactor accident.

8. Chernobyl

Nowadays, we know better the nature and extent of radioactive doses released to the atmosphere, thanks to measurements made in the USSR and in the several countries affected by radioactive products after the accident (26 April 1986, at 01:23 a.m.). The results of these analyses have been communicated to the IAEA-state members.

The radiological-protection actions taken in the USSR and in other countries speak very clearly of the size of this nuclear catastrophe; large evacuation of populations in the accident area; a concrete coffin for the damaged reactor; and massive elimination of irradiated foods (milk, meat, vegetables, etc.). 115,000 people were evacuated from Chernobyl and its surroundings. Radiation measurements made there demonstrated that maximum doses, to which some critical population groups of the city were exposed, may have reached 0.1 Gy for external irradiation and approximately 1 Gy for beta rays at the skin level. The International Radiological Protection Commission has recommended that populations should be evacuated when the radioactivity level is larger than 0.5 Gy. Apparently, the efficient actions taken by authorities in the USSR have shown that, in most cases, the doses in local populations were lower than 0.3 Gy, and were 0.3 and 1 Gy in 2% of cases, but reached 0.17-0.24 megabecquerels (Mbq) in some children between three and eight years old.

Dosimetric analysis made as a consequence of the Chernobyl accident by the IAEA laboratories at Seibersdorf give a real idea of the size of this nuclear accident which took the lives of at least 30 people. The tests made from 29 April to 30 May 1986 at Seibersdorf on samples of air, rainwater, and fruit have established the presence of unprecedented concentrations of radio-nuclides (barium 140, cesium 134 and 174, iodine 131, 132, and 134, molybdenum 99, ruthenium 103 and 106, strontium 90, tellurium 129 and 132, cerium 141 and 144, niobium 95, rhodium 106, ruthenium and zirconium 95). The air analyses showed the activity, above all, of tellurium 132 and iodine 132 to a maximum value of 91 Bq. per m^3 on the first of May 1986 and a maximum of 0.04 Bq. per m^3 on 10 May. For iodine 131, the maximum activity was 132 Bq. per m^3 and increased 8% the next day.

The enquiry directed by the Academy of Sciences of the USSR pointed out that a human error was mainly responsible for the Chernobyl accident. The lessons of Chernobyl pleaded for tightening nuclear-safety norms almost everywhere reactors exist. The other benefit from Chernobyl has been to put in place, under the control of the IAEA, an accident information system that allows the instant notification of other countries. A sincere will of multilateral assistance was demonstrated to help the victims of that nuclear catastrophe. Three Mile Island and Chernobyl have shown that major nuclear accidents may always occur, despite continuous reinforcement of nuclear safety norms.

9. Conclusion

Because nuclear energy is a high-risk technology, it is not possible to appease public opinion on the reduced probability of major nuclear accidents. The more one builds reactors, the more one increases the risk of major nuclear accidents, even though the probability of their occurrence is very small. It is generally agreed that forecasts made by planners, public or private, on the development of nuclear-power plants for the year 2000 must be reduced. Nuclear energy has come to stay forever, despite the controversies and risks that accompany it. There are numerous countries that cannot satisfy their energy deficit other than by going nuclear.

The accidents at Three Mile Island and at Chernobyl have added to public mistrust of nuclear activities. It is a fact that, to a very great extent, public opinion underestimates the gigantic efforts made by nuclear countries to find dependable solutions to the problem of radioactive waste. France, for example, is planning to start the operation of two installations at La Haye for the vitrification of radioactive waste, each capable of producing 90 kg per hour of borosilicate glass. In spite of all these technical efforts, radioactive wastes pose an ethical problem which will probably not have a solution. The long radioactive life of many nuclides, and mostly the extreme danger of some of them, are responsible for this situation. Do we have the right to bequeath to future generations the consequences of energy choices for which they are not responsible? It is because of such questions that the international nuclear community must look for global solutions together with governments and with the public.

In the case of Zaire, my country, we have signed the Non-Proliferation Treaty. After a long time, we also have made the wise decision neither to enter into the nuclear arms race nor to construct NPPs. (Zaire has an experience of more than 30 years of nuclear research.) And concerning nuclear wastes, it is either an irony or a blessing, that Zaire exports them to the developed countries.

SECTION TWO

Key Policy Issues: Energy Choices of Developing Nations, Nuclear Proliferation and Global Warming

The Impact and Safety of Commercial Nuclear Energy: Perspectives from the Philippines

Achilles del Callar
Adamson University
Metro Manila, The Philippines

1. Present Status: Mothballed

In a paper presented to the Radioisotope Society of the Philippines on 28 February 1989, Fidel S. Correa, Manager of the National Power Corporation (NPC), Operations Division, claimed that the Philippine Nuclear Power Plant (PNPP) construction was 98% complete, and that pre-operation testing (pre-fuel loading) had been completed. Mr. Correa is now in charge of mothballing the PNPP with the objective of keeping "core load readiness," at an annual cost of about $10 million (US).

On 21 April 1987, the Philippines' national government signed a "Memo of Agreement" with the NPC, transferring from the NPC to the national government P33.7 B (about $2 billion (US)) of "assets" and P29.6 B (about $1 billion (US)) worth of liabilities of the PNPP. What was disturbing about this memo was the needed and given consent of all foreign creditors involved with the financing of the PNPP. The basic loan of $644 million (US) came from the EXIM Bank in 1975. Every single loan thereafter, mostly from consortia of banks, had cross-default conditions. Further, failure of the country to stay in the good graces of the International Monetary Fund (IMF) constitutes, in every loan, a cause for default.

As of 1989, the external debt due to the PNPP had ballooned to $2.6 billion (US), despite the fact that $1.2 billion (US) has already been paid by the Philippines in interest and capital, as of May 1988. The only other available information about this debt is the fact that $355,000 is being paid out daily on this single biggest loan of the country. Half of the 1989 national government's budget is allocated for external debt payments (about $5 billion (US)).

For a nation severely buffeted by population stresses - the country's population grew from a mere 18 million in 1945 to 62 million in 1988 - this daily $355,000 US toll equates to 37,000 housing units per year, or 33,000 classrooms, or 2,100 km of two-lane, asphalt roads, or 7,000 jobs. The housing shortage is so acute that about 2 million metro Manila residents live in substandard housing, mostly packing-crate shanties. About 70% of the total population is living at or below poverty levels (now pegged at P4500 monthly income for a family of six, or about $215 (US) monthly income. In early September 1989, more than 450,000 public-school teachers staged walkouts in a demand to have their salaries increased to the poverty level of P4500 (about 215 (US)) per month. The demand, which was not granted, would have cost P5 billion annually, or less than 5% of the $5 billion (US) allocated and paid annually for the external debt.

2. A History of Bribery and Corruption

Even at the bidding stage, questionable transactions dogged the PNPP. General Electric submitted specifications for two 600 MW plants at a cost of $400 million (US). Westinghouse turned in a brochure for the same bid at $500 million (US). Westinghouse "won" the bid with the assistance of "agent" Hermino Disini, husband of the personal physician and first cousin of Imelda Marcos. Anomalies did not stop here. Disini's group of companies, HERDIS, which never had any construction experience, set up "Power Contractors, Inc.," and won the bid for civil-works construction of the PNPP. Still another "instant" HERDIS company, Summa, cornered the PNPP insurance, despite government regulations that government projects be insured with the Government Insurance System. Yet still another instant HERDIS creation, the "Technosphere Consultants Group" (TCG), received commissions from Westinghouse on work purportedly done on the PNPP project. Oddly enough, all payments made by Westinghouse to TCG were made abroad, reinforcing local speculations that all such payments were mere bribes.

In a year's time, the cost of the PNPP, with the intervention of the US State Department, rose progressively from $500,000 million (US) to $800,000 and then to 1.1 billion (US) -- at which point the price suddenly became not for two but for just one reactor. As then Commerce and Industry Secretary Vicente Paterno quipped: "We are buying one reactor for the price of two." A few months later, Secretary Paterno resigned from the government.

3. On Nuclear Safety

Construction of the PNPP began in 1976. In June of 1979, Marcos convened the Puno Nuclear Power Hearing Commission, ostensibly to investigate the safety of the PNPP in the light of the Three Mile Island (TMI) accident. The handful of nuclear power opponents jumped with joy at the chance of having a public avenue to vent their concerns on nuclear safety. Little did they know that the real reason for calling a hearing was to find an excuse to stop the construction of the PNPP, so as to give Westinghouse time to get an export license for the reactor. At that time, the US Nuclear Regulatory Commission (NRC) was insisting on reviewing the design of the PNPP. Westinghouse was just as intent on evading this NRC review. One can only surmise that the PNPP would not have passed US NRC standards. This appears to be the cause of Westinghouse's evasive actions. Three months after the initial hearing, the Puno Commission declared the hearings closed and recommended the construction of the PNPP stopped. Marcos immediately ordered implementation of the recommendation. In May 1980, Philippine Solicitor General, Estelito Mendoza, hired a US lawyer and presented the US NRC with an "amicus brief" stating that the "safety of the

Filipino people is the concern of the Filipino people, and that review of the PNPP by the US NRC would constitute a violation of Philippine sovereignty." This statement came from a country with hardly any personnel qualified to evaluate a toy reactor! Thus it came to pass that a US NRC memo was sent to the US State Department, declaring that "it would not evaluate this particular nuclear reactor, lest it violate Philippine sovereignty." The US State Department then issued Westinghouse its coveted export license. With the export license secured, Marcos announced that a new contract had been renegotiated with Westinghouse, and that construction of the PNPP would resume immediately.

The Puno Commission had barely enough time to tackle the issue of site safety. The sudden decision to close the hearings prompted a dramatic walk-out of the single Opponents Panel which made life hard for 17 government panels, NPC, Westinghouse, and the Ebasco panels. The PUNO Commission came out with reasons (never taken up during the hearings) for stopping the construction of the PNPP: 1. The PNPP is not safe. 2. No permanent burial site for nuclear waste has been found. 3. To secure the safety and operate the plant would entail future costs that are both uneconomical and beyond the means of the Philippine government. Regarding safety, the Puno Commission made the following points.

- The PNPP, as designed, is not safe. Admittedly, it is an old design, circa 1972, plagued with unresolved safety problems, like other Westinghouse designs under review by the US NRC. Thus the PNPP is a potential hazard to the health and safety of the public. (It may be noted here that the PNPP used blueprints for the Yugoslavia reactor, as well as unshielded cables that were rejected by Korea.)
- The PNPP design needs fundamental changes and additional safeguards. It appears that Westinghouse nuclear reactors do not have, among others, adequate emergency core cooling systems.
- The frequency of accidents in nuclear power plants, not excluding those designed by Westinghouse, are ominous signals that safety is not assured and, therefore, that additional safeguards are imperative. Express warranty on the part of the supplier as to the safety of the plant, including all its vital components, should be required.

The renegotiated contract did not require Westinghouse to comply with US fire-protection requirements to guard against accidents that have been caused by fires in US Westinghouse reactors. Neither did the new contract contain any provision that would lead to the resolution of even a single unresolved safety issue. Of the 44 new safety requirements on Westinghouse reactors, as a result of TMI, less than half were incorporated into the new contract.

Another serious area of concern was the lack of funds, equipment, and personnel of the Philippine Atomic Energy Commission in checking the compliance of Westinghouse with nuclear safety requirements. PAEC, now renamed the "Philippine Nuclear Research Institute," cannot even fix the swimming-pool-size leak of their recycled TRIGA research reactor. The leak, discovered in April 1988, has yet to be repaired, not only due to lack of funds, but also due to lack of competence in very basic engineering skills.

The Preloading Audit conducted by the Operational Safety Review Team of the IAEA in 1985 found at least 4,000 deficiencies. This was not surprising, because 48,000 field changes were made during construction. A vital nuclear-safety component, the steam generator, was dropped during unloading at ship-side. It was dropped again, on site, where a steel rod penetrated the generator. Repair was done on site, with hammering to remove the dents, and welding to fix damaged tubes. Such tubes were fabricated under very complex annealing processes.

Other safety deficiencies surfaced during the Licensing Hearings conducted by the PAEC in 1985. These included the flooding of the reactor-vessel building during heavy rains, causing water to flow out of the control panel of the sea-water loop; arcing of unshielded high-voltage cables; and cracking of turbine blades during pre-run testing. Again, as in the case of the steam generator, the cracked blades were welded on site. At the PAEC Hearings (a "kangaroo court," if ever there was one), Westinghouse "experts" joined experts from NPC and MIESCOR in explaining the foot-high flooding of the reactor-vessel building as resulting, not from rain seepage from outside, but from water coming out by "sponge action" from the drying concrete. High-voltage arcing was claimed as part of normal operating procedures. Scattered throughout the plant were numerous pipe hangers, whose base plates were welded to extension plates. This was done because screws in the base plates would not hold, due to bad concrete.

Lack of safety valves and pressure gauges caused an auxiliary water tank to implode when water was pumped out too fast. A favorite non-compliance story related by the late PAEC Commissioner, Zoilo Bartolome, concerned two pipes which were interchanged in position. One was painted green, the other orange. Rather than interchange the pipes, Westinghouse had the orange pipe repainted green, and the green pipe orange. This incident reflects the rather lackadaisical attitude of Westinghouse towards nuclear safety.

Interwoven with safety concerns is the country's lack of sufficient industry to support a nuclear-power plant. Any replacement part will have to be flown in from the US. If this part happens to be needed to prevent an accident, then BOOM!

4. The PNPP Future

On 1 December 1988, the Philippine Government filed a suit against Westinghouse Electric Corporation and Burns and Roe Enterprises, for bribery and corruption in negotiating the PNPP contract, and for delivering a technically unsafe plant. The suit did not fare too well. The New Jersey District Court remanded the charge of technological deficiencies to arbitration, as specified in a contract which the Philippines claimed was null and void because it was obtained through bribery and fraud. Nevertheless, the New Jersey court decided to pursue only the charge of bribery and fraud against Westinghouse and Burns and Roe.

An arbitration procedure seldom gets any publicity. No details are available on its progress. Rumor mills have it that Westinghouse is willing to complete the plant, at its own expense, provided the reactor is allowed to operate. The Philippine Nuclear Power Plant saga is not yet over. Time will come when IMF pressure, because of the country's external $30 billion US debt, will force the country to run the plant. Even the government is struggling to survive, due to this debt. The political will which closed the PNPP may no longer exist in a year's time. The PNPP is just hibernating.

5. The Future of Nuclear Plants

Nuclear-power-plant manufacturers are now stressing "passive" safety design features. Simple gravity flow, for example, forms the emergency-core-cooling system. Gas-cooled versions are supposed to remain structurally intact, even with total coolant loss. Others favor liquid-sodium coolants which can operate at higher boiling points, absorb unexpected temperature surges, and require no pressurization. (Sodium reacts violently with air or water.) A SEA BROWN BOVERI of Sweden proposes PIUS, "Process Inherent Ultimate Safety," a water-cooled reactor surrounded by a large pool of cold, heavier, borated water. Sheer pressure of the reactor coolant is supposed to separate the two fluids. Loss of reactor coolant would cause the heavier fluid to flow in. No prototypes, however, have been built.

Other approaches include standardized designs to bring down costs. Westinghouse says 600 MW models are safe sizes. General Atomics is using 135 MW modules constructed underground which can be installed and operated in clusters of four. Helium-cooled proponents point to the Hamm (Ruhr region) 300 MW reactor which has been on-line since September 1986.

It must be recalled that nuclear-reactor proponents in the 1970's were mouthing that nothing, but nothing, could happen in nuclear reactors due to their "defense-in-depth" safety design. This suggests that the present propaganda of manufacturers, taking advantage of the Greenhouse Effect, must be taken with a grain of salt. Nuclear engineer Lawrence Lidsky of MIT says, "The water-cooled reactor in any incarnation is inherently an exceedingly complex, unforgiving device." Lidsky runs gas-cooled reactor research at MIT. An engineer, at the Cap de la Hague reprocessing plant in France, has this to say to Robert Jungk (author of *The Nuclear State*, 1979): "Inevitably the plants will become so faulty, one after another, they will have to be closed and walled-up and turned into radiating monuments to our 'progressive' age. There will not be a catastrophic explosion but this region will be turned into an atomic wilderness. It will be so polluted that in a few decades the whole cape will have to be hermetically sealed and written off as a dead loss....Then the last human being will leave Cap 'Coul'

Hague. It will have to be guarded like a seat of infection for decades and centuries by people who will not understand why we have permitted this. People will hate us." At least he is talking of sealing off the plant and region. In 1975, the West-Valley reprocessing plant in New York was simply abandoned by its owners who thought it was good business to do so rather than spend $600 million (US) to meet safety standards when the plant only cost $30 million to build.

6. The Future Hinges on Radioactive-Waste Storage

April 1989 marked the tenth anniversary of the TMI accident. The clean-up bill has gone up to $973 million (US) - about seven times the original estimate. Another $20 million will be charged for the storage of 150 tons of TMI radioactive debris until the Federal Repository at Yucca Mountain, Nevada, becomes operational. More dollars will have to be spent decontaminating the water that has been sitting in the reactor vessel at TMI. The message from TMI is clear: the problem of radioactive-waste storage will not disappear any time soon.

Still another message comes out of TMI. There is more to learn about nuclear reactors. Research on TMI, and now Chernobyl, has been extensive. The November 1988 conference of the American Nuclear Society focused on the TMI accident alone. The US NRC has an entire room on TMI literature. Yet engineers still don't understand why Unit 2 didn't melt down, why the outer wall held.

What happens to radioactive waste? The stockpile of high-level radioactive waste grows by the tons, almost all in temporary storage. At least 11,000,000 cubic feet of liquid waste from US reprocessing plants are stockpiled, mostly in Hanford, Washington. By the year 2004, when the US Department of Energy (DOE) will have its first permanent storage site, there will be 53,000 tons of commercial spent fuel. There is little choice between storing liquid waste (which is harder to store because of its acidity and liquid form) and storing spent fuel which, although still protected by its cladding, can form critical masses.

Nuclear power, with all its daunting problems, is again being promoted because of the Greenhouse Effect. Radioactive waste is a nightmare that will not disappear when one wakes up. Nuclear power is unsafe, expensive, and with pollution effects which will last through nigh all generations. The biblical phrase, "from henceforth all generations will call me blessed," might well turn into "from henceforth all generations will call me accursed," for having left a legacy of radwaste.

Setting the Problem of Nuclear Energy in the Developing World Context

C. Araoz
Comisión Nacional de Energia Atómica
Buenos Aires, Argentina

1. The Need for Energy

For the discussion of developing-world opinions, attitudes, and actions in relation to nuclear energy, it is important to consider their points of view, philosophies, and experiences. These generally are quite different from related views in developed societies. Some of these points are recognized by the world community and are included in the concept, formulated by the World Council of Churches (WCC), of a "just, participatory, and sustainable society." For instance, there is general agreement that the difference in economic development between industrialized and developing countries has to be reduced. This is an important step on the way to a just society. If we want to increase the gross national product (GNP) in developing nations, then a proportional increase in electricity generation is required. In the case of Argentina, a 4% annual increase in the GNP requires a 6% annual increase in electricity generation. If we wish to achieve a worldwide just society, this will require a fair distribution of energy.

Present and expected world population growth in industrialized countries is one-half to one-third of the rest of the world. If we consider that in a better world, the gap in per capita energy consumption, among different nations, must be reduced, more energy sources have to be available in developing countries than in industrialized nations. This means that, if present annual electricity consumption, per capita, in industrialized countries is 7000 kwh, and 500 kwh in the rest of the world, to reduce this gap, for example to one-third, some million Mw(e) should be installed.

We could say that, with some exceptions, to implement this mix of energy consumption, nuclear energy in the non-industrialized world is not urgently needed, and today's needs could be satisfied with thermal or hydro plants. Nevertheless, nuclear programs have to be started, sometimes including one or two nuclear-power stations, in order for developing nations to be acquainted with the technology for the future. We have to consider that 10 to 15 years are necessary in developing countries (DC) between making a decision and the implementation of the required infrastructure. International organizations like the International Atomic Energy Agency (IAEA) play an important role in advising on required steps for nuclear programs.

It is interesting to remark that, nowadays, the world average for nuclear electricity participation, in total generation, is 15.5%. Nuclear electricity generation contributes about 3% to electricity production in developing countries, compared with 18% in developed nations.

According to the International Atomic Energy Agency (IAEA), nuclear generation is expected to have a small percentage increase in the next 10 to 15 years, even with its present unfavorable situation. This suggests that nuclear generation and related activities could be expected to increase in developing countries. One reason for this increase is that a mix of energy sources is important to have the necessary flexibility for future changes and to cope with unforeseen events in resource availability.

Energy planning is long-term planning, for which adequate information and policies are not always available in developing countries. Such plans have to include not only the needs of the inhabitants, but also industrialization policies, structural changes in the economy, and so on. As a result, we should be careful with electricity-demand projections, because they have often proved wrong in the past, both in their estimates of demand and of the best mix for a given electrical power system. As a consequence, the need for energy in DCs has to be carefully considered in energy forecasts. In Argentina, we have always had electricity shortages and blackouts, for the last thirty years.

Energy needs in the developing countries have been the subject of many studies. During 1980-1982, three consultations were organized by the WCC in Asia and Latin America. Their purpose was to analyze the ethical, social, and technical implications of energy developments, within the context of "just, participatory, and sustainable energy policies for development." In September 1987, the Glion Working Group of WCC had a meeting where nuclear power and the energy needs of Third-World countries were analyzed. Last year, the IAEA issued a technical document on "Energy and Electricity Demand Forecasting in Developing Countries," where possible scenarios of socio-economic and technical developments were taken into consideration to forecast energy needs. I suggest that WCC verify if the document is applicable to WCC ideas for a just, sustainable society, and make their comments to IAEA.

2. Policies and Requirements for Nuclear-Energy Generation

Among the many requirements for nuclear electricity, several are of particular interest in developing countries. For one thing, DCs must be prepared to define, for themselves, their energy policies. They have to develop capabilities to perform the analysis of energy options within the overall context of their economic development and long-term policies. This cannot be left in the hands of foreign consultants. Government commitment to a given energy choice has also to be defined. Capabilities for financial planning (appropriate tariff structures, project financing, etc.) in government and utilities have to be developed. There are several models for implementing a nuclear-energy program. The decision and choice for the model that better fits a country's needs is a political decision, and not the result of scientific methodology. The model that ought to be chosen is one that is able to avoid energy dependence in the future.

Ideas of independence have deep roots in the non-industrialized world. The desire to be independent, in important aspects like energy, is often difficult for developed countries to understand. Independence has played an important role when decisions are made about alternatives to implement nuclear-energy capabilities. For instance, we have chosen natural-uranium reactors just because of independence in fuel supply, though this type of reactor produces electricity that is approximately 10% more expensive than enriched-uranium reactors.

Another requirement for nuclear electricity is advanced technology. This requires qualified personnel able to acquire, adopt, or develop such technology. Long-term planning and continuity is essential to achieve an adequate level of technical development. This could take between one to two decades, depending on the country. In the case of Argentina, after 10 years of our nuclear program, we knew how to buy a nuclear power plant (NPP), how to deal with technology transfer agreements, how to start the implementation of a nuclear industry, and so on. With the second NPP, and another 10 years, we are beginning to see what the nuclear business is like.

To achieve a "record of excellence" for nuclear generation, countries must also develop support for nuclear-plant operation. This includes technical support from the national nuclear organization, nuclear and industrial services, and so on. Likewise, an assured nuclear-fuel supply is essential to the operation of the plants. Spent fuel must be disposed of, corresponding options analyzed, and programs started in due time. Nations must also develop national legislation for radiation protection and nuclear safety and establish regulatory organizations. Although there is extensive international knowledge available in this respect, in some developing nations, decisions for effective implementation are not taken in time. Finally, "nuclear nations" must develop a participatory society for the discussion of energy needs and options.

3. Strategies and Options

Many times, successful societies and good energy models have shown the way to other countries. Today, comparisons are inevitable, and these increase with time, as communications are better and faster. Consequently, events and policies of industrialized countries have an impact on the developing world, and many are implemented in their national policies and attitudes, but with some years' delay. This happens also with the nuclear controversy and the present discussion of energy options.

History shows that in many cases, political independence was achieved after much fighting for freedom. Then comes economic independence, energy independence, technology independence, development of business possibilities, and so on. These facts have to be recognized by industrialized countries, and consequently they have to solve and clearly define their fair, or reluctant, attitude toward independent development of other countries. Nuclear-development strategy, in the context of an energy-independence policy, plays an important role in this situation.

Several developing countries, including Argentina, have taken the option of a "conveniently independent nuclear program." "Conveniently independent" means that the investments for nuclear developments are not carried on at any cost, but under well defined economic limits. For instance, electricity generation cost (mills/kwh) is an important factor for decisions. Another factor is that participation of local industry in the installation of a NPP has a limited higher price allowance when offers from local industry are compared with those from foreign manufacturers. Countries like Argentina must also analyze the impact and applicability of nuclear technology in other industries and their economic consequences. When convenience has been demonstrated, technology-development costs are analyzed, and decisions on investments are made. Decisions for assurance of supplies include cost-risk considerations. This "conveniently independent" approach indicates a clear difference between peaceful nuclear programs and military applications. In the military applications, economic aspects are not considered relevant.

Some nuclear programs of developing countries include strategies to promote the introduction into local industries of high technology, procedures, quantitative analyses, and so on. Government organizations and vendors have taken such strategies as an important part of decisions on different energy alternatives. This goal of many nuclear programs (to develop local-industry capabilities, engineering expertise, pilot plants, laboratory installations, technical cooperation programs, and so on) is not typical in thermal and hydro-electricity cases, a situation that could be modified in the future.

Assurance of supplies for spare parts, technical services, components, fuel, and so on, also plays an important role in energy decisions. The reason is that, when the capital investment is already made for a NPP, three decades of uncertainties, regarding supplies, begin for the utilities. Today, if an international supplier is not considered reliable, because of political, commercial, or technical reasons, he will never get a purchase order. There is also a remarkable difference on this point among thermal, hydro-, and nuclear-power facilities. For developing countries, larger dependency means larger uncertainty.

Some developed countries have made the decision to stop installation of nuclear power plants. This type of nuclear moratorium will have different effects in industrialized countries than in the rest of the world. In the case of DCs, it would take one or two decades to build up nuclear capability again, after a moratorium. In the case of industrialized countries, the build-up time is very short, and a moratorium has no influence on their capability to perform and implement policy changes. This difference is important because a nuclear moratorium will define a DC as a future "turn-key" buyer of all kinds of nuclear supplies and services.

4. Conclusions

Alternative sources of energy (solar, wind, geothermal) have to be developed, although they cannot be expected to satisfy the large energy needs of the developing world. The environmental impacts of world energy consumption and its undesirable effects are, to a large extent, created by the energy needs of industrialized countries, who now have approximately 70% of the world energy demand. Significant changes in this situation are not expected in the next 15 years. Consequently, ethics demands that the wide use of alternative sources of energy, and that a reduction in per-capita energy consumption, claimed by some organizations, start in the industrialized countries, not in developing countries. Instead, DCs need to promote better living conditions in their own nations.

Improved calculation models to forecast energy needs in non-industrialized countries have been developed by the IAEA. These include socio-economic and other impacts. Regardless of whether such models may be in agreement with the WCC points of view, profitable discussions could take place.

It can be expected that nations of the developing world will always look for an independent energy policy. Many of them have started nuclear programs, and many will start them in the future, so as to be prepared in case the nuclear-energy option is needed in the medium or long-term future. The task for energy planners is to determine the best mix (nuclear, coal, oil, gas, hydro and renewables) for growing energy needs, at minimum costs, with an acceptable independence and environmental impact. Polarization and emotional controversy, regarding the nuclear-energy debate, has to be prevented. Government should provide information to the public, on energy matters, at a very early stage of the planning process. This is important, because decisions among energy options involve complex procedures; items of different natures, and different groups of opinions and interests, must be included. Perhaps a procedure has to be developed to organize the public discussions. The need for analysis, by the different interested parties, has to emphasize the ethical aspects involved in energy decisions.

A nuclear-weapons program is definitively a political decision. It is not the consequence of a technical capability in nuclear energy for peaceful uses. This is shown in almost all cases of industrialized and developing countries. The tendency of many developing countries to start nuclear activities and to have sustained developments in commercial nuclear electricity generation is based on the above-mentioned points. Considering their particular characteristics, which in many aspects differ from those of industrialized societies, it is important to evaluate the possibilities that a peaceful nuclear program could be oriented to a nuclear-weapons program. For this reason, the objectives which are behind independent nuclear developments, many of which are given here, need to be carefully considered. A political decision in favor of a nuclear-weapons program could be made by a developing country at any time, before or after having the corresponding nuclear capability. To have this nuclear capability has not been the determining factor. We all know of developing countries that have non-peaceful nuclear programs and they have not yet installed the first nuclear-power reactor. Likewise, many other countries only have commercial applications.

Our desire to be independent in future energy generation is the reason why we have been accused, during the last 30 years, of having the bomb "in the back of our minds." During this period, we have changed many governments, with different political parties, with non-democratic military rulers, and we have even had a war. Given all these factors, it might be more important to dig deeper into the subject, to analyze why Argentina never changed its political decision to have a sustainable, peaceful, nuclear program. I hope I have presented some points for this analysis.

References

"Necessita dell energia atomica e del uso pacifico: intervenio della Santa Sede Alla XXVI Conferenza Generale dell' AIEA," Vienna, IAEA.

"Promotion and Financing of Nuclear Power Programs in Developing Countries," Vienna, IAEA, Senior Expert Group Report, 1987.

"Nuclear Power: Status and Trends," Vienna, IAEA, 1987.

"Energy and Nuclear Power Planning in Developing Countries," Technical Report Series No. 245, Vienna, IAEA, 1985.

"Cost and Financing of Nuclear Power Programs in Developing Countries," TECDOC 378, Vienna, IAEA, 1986.

"Experience with Agency's WASP for Nuclear Power Planning in Developing Countries," TECDOC-364, Vienna, IAEA, 1986.

"Highlights from the XVIII Council of Experts and the XIX Meeting of Ministers," Latin America Energy Organization (OLADE). *Energy Magazine*, 12, No. 3 (December 1988).

"Manpower Development for Nuclear Power," Technical Report Series No. 200, Vienna, IAEA, 1980.

H. E. Bocco, "The Argentine-Brazilian Nuclear Cooperation," Facultad Latinoamérica de Ciéncias Sociales Programa Argentina, (FLASCO), June, 1989.

Faith and Science in an Unjust World, WCC Conference on Faith, Science and the Future, Vol. 2, 12/74, MIT, Cambridge, USA, July, 1979.

Candu Update, Atomic Energy of Canada (AECL), AECL CANDU *Operations Communications*, September 1, 1989.

Stanley Hatcher, President, AECL, *Ascent* 10, No. 2 (Summer 1989).

A. T. Rapanelli, National Director, Energy Conservation and New Sources, *Ciencia Nueva* (April-May 1989).

Issues and Experiences Concerning Nuclear Energy and Nuclear Proliferation

B.C.E. Nwosu
Research and Technology Limited
Lagos, Nigeria

1. Introduction

The work of the Church and Society Subunit of the WCC on the nuclear energy topic has been almost prophetic and certainly a significant contribution to ecumenical thought. As a nuclear physicist, it has been a learning experience for me to participate in WCC meetings. The discussions also have given me ample opportunity to present the Third-World perspective, particularly the African perspective. Thus, it is my view that I can contribute to the learning experiences of my colleagues from the highly industrialized countries. We cannot have a proper understanding of the nuclear-energy issues, however, without a brief overview of the history of the development of atomic energy.

2. Historical Background

Located at the west end of the University of Chicago Football Stadium are these words:

> *On December 2, 1942, man achieved here the first self-sustaining chain reaction and thereby initiated the controlled release of nuclear energy.*

It was Enrico Fermi and his colleagues at the University of Chicago who recorded for history these simple, but dramatic, words. This statement was made in December, 1942. On July 16th, 1945, the first atomic-bomb test was made on the sands of New Mexico, and **three weeks** later, the first atomic bomb was dropped on Hiroshima. Two days later, another bomb was dropped on Nagasaki, Japan. With these events the course of history changed permanently. Please note that only three years elapsed between the control of nuclear fission and the manufacture and dropping of the atomic bomb on Japan! Arguments have raged on whether the United States should have used this weapon to subdue Japan. I, myself, have jumped from one side of the fence to the other. I recall a visit in 1960 to Wright-Patterson Air Force Base, in Dayton, Ohio. After seeing the Japanese hari-kari suicide airplane, I felt that only something like the A-bomb could have ended the Second World War. Ten years later, in 1970, as an inhabitant of defeated Biafra, I did not feel quite the same.

It is very difficult to trap a lion, alive, in our African jungle. But this can be done. It is much easier to cage the lion, and at an opportune moment, release it to ravage an enemy's compound, than to tame the same lion and make it perform household chores. In this simple illustration, most of the major issues of nuclear energy can be summarized:

- How do you capture and cage the lion? That is, how do you acquire the nuclear technological know-how?
- How do you release the lion to your enemy's compound? That is, how do you deliver the nuclear bomb?
- How do you ensure that the lion, after eating up your enemy, does not turn round to eat up you and your family? That is, how do you solve the attendant problems in atomic fallout and pollution arising from nuclear explosions spreading and causing havoc world-wide?
- Once people know that it is possible to use a lion to deal with an enemy, is it right to restrict the knowledge only to a few persons? That is, what moral right has anyone to deny other countries the right to acquire nuclear weapons?

And finally, we now leave the lion analogy alone. Nuclear fission has been harnessed. It has been domesticated. Although it was initially used for military purposes, now atomic energy is being used successfully for peaceful purposes in agriculture, medicine, industry, and so on. The most significant application is in generating electricity. These peaceful uses have raised many problems pertaining to economics and world-consumption patterns, nuclear-waste disposal, and above all, safety. Many issues are interrelated. How do they relate to Africa and the Third World?

3. Acquisition of Know-How

I deliberately chose the expression, "know-how," to avoid the controversial term, "technology transfer." The single, most pressing issue of nuclear energy, as far as African countries, i.e., black African countries, are concerned, is how to acquire nuclear technology for peaceful uses. The problem is complicated because of the military factor. Any time a less developed country shows interest in nuclear energy, questions are raised as to what that country really wants. Is it for peaceful use, or is it to make the atomic bomb? Questions may be raised about the economics, but really, what they are worried about is the bomb. We shall see later, when we discuss the issue of nuclear proliferation in greater detail, that these doubts and fears are not entirely without foundation. The main obstacle to African countries' acquiring the nuclear know-how is from the big powers, who already have huge arsenals of atomic bombs, as well as enormous nuclear-power plants.

The problem is that many African countries do not really know what they want. The concept of self-reliance is floated and bandied around. Every university has a department of physics which offers courses in nuclear physics. For decades, some countries, like Nigeria and Ghana, have had laws establishing Atomic Energy Commissions in their statute books. There is nothing significant to see in these countries, yet we Africans must acquire the nuclear know-how for peaceful uses. If, for no other reason, nuclear science is an important part of human knowledge. All human knowledge that is good comes from the Almighty. For another reason, when all the nagging issues concerning the numerous uses of nuclear energy, such as waste disposal, have been resolved, we do not want to discover that the train has left us, as it did in the Industrial Revolution. What should African countries do so as to acquire peaceful nuclear energy? I have several insights.

- Self-reliance is not self-sufficiency. Each African country or group of African countries, such as ECOWAS, should identify a friendly country, or group of countries, with which to collaborate.
- Resources should be pooled together, however small, to make a reasonable start.
- A choice of a suitable base university or institute within the country or region should be made.
- Workforce development, using the resources of the IAEA, as well as those of friendly countries, must be executed reasonably and resolutely.
- Projects should be mission-oriented and integrated within the overall national or regional, political, economic, and social plans.

To sum up, African countries cannot do it alone, nor can they be taken seriously if they themselves do not do the necessary homework by themselves. They must acquire nuclear technology for peaceful uses in a realistic manner, or condemn their children and their children's children to be fourth-class citizens of the world.

4. The Nuclear Bomb

Africa does not need a nuclear bomb. My proposition is not based on religious or moralistic grounds. Events in the last two decades have overtaken any such aspiration by African countries. The level of technology, of nuclear armaments and delivery systems achieved by the superpowers has made such an aspiration simple madness. What, then, is to be done about South Africa and Israel? A majority of African countries, rightly or wrongly, consider Israel and apartheid South Africa their enemy. There is no doubt in my mind about South Africa. However, I do not agree with the OAU position on Israel. It is now known that these two countries have nuclear bombs, or at least the ability to make them at short notice. While the delivery capability of Israel is not in doubt, that of South Africa is, in my opinion, in doubt. Their bluff can be called, if African countries are really serious. What should we do?

- Black African countries should recognize Israel. Israel has returned Egyptian territory, i.e., African territory, to Egypt. Egypt has recognized Israel. I see no more reason why black African countries should push Israel into the arms of their enemy, apartheid South Africa, by trying to isolate Israel diplomatically.

- African nations should establish just, competent, and fair governments. Radio South Africa relishes in broadcasting sometimes-exaggerated stories, on apartheid-type actions, which African governments take against their fellow citizens or neighbors. These actions by our governments reduce our capacity to obtain full support from the international community.

- The African economy is in shambles, due to massive mismanagement of resources and corruption aided and abetted by foreigners who have easy access to nationals willing to front for them. We must feed our teeming population, using modern, appropriate technology. The greatest enemy we have is hunger, poverty, disease, and ignorance, and not the South-African bomb.

- Having done all of the above, it is quite easy to isolate South Africa diplomatically, and even economically, without fighting a conventional or nuclear war.

Summarizing, I maintain that, while African countries need nuclear energy for peaceful purposes and in order to improve their economies, they do not need any nuclear bombs. They can contain their enemy, apartheid South Africa, simply by putting their house in order, politically, economically, and ensuring social justice for their citizens.

5. Nuclear Power Plants

The best example of the domestication of nuclear energy is the nuclear-power plant, now widely used in the highly industrialized countries (HICs) for the production of electricity. The development, in commercial quantities, of nuclear-power plants came much later than the development of the atomic bomb. It is considerably easier to produce the atomic bomb, even if it is the so-called "dirty bomb," which South Africa claims it has developed, than to develop a nuclear-power plant with all the safety precautions that are built into it. Once again we observe the parallel between harnessing nuclear energy and taming the lion.

Given the situation in the HICs, such as the USSR, and the handling of the Chernobyl accident, one is tempted to ask, "why should any African country be interested in establishing a nuclear power plant?" Professor Kosta Tsipis has reviewed in one of the WCC background papers, the issue under the heading, "Nuclear Power and Energy Needs of the Third-World Economies." Professor Tsipis states that "the decision to go or not to go nuclear, and if yes, what kind of technology a Third-World country should adopt, is clearly based on current and future energy requirements of the Third-World country and on the kind of energy development that a Third-World country has so far pursued."[1] This means there must first be some homework. Also, a look at the IAEA guidelines for nuclear-power introduction and implementation demonstrates how very difficult it is for the less developed countries, such as Nigeria, to undertake the nuclear-power plant project. Nevertheless, my position is that, much as it is extremely difficult, given the manpower position, the economy, and the technological infrastructure, the less developed countries (LDCs) should not foreclose the nuclear-power option.

I maintain that the so-called energy "soft path" option, such as solar energy, which happily is quite popular in Nigerian academic circles, is still a long way from making a significant contribution to the LDC's energy needs. It is clear that we still need to have a conscious, well articulated energy policy, as well as research and development, and above all, the political will to implement approved programs. Also, the gap between the research and development efforts in solar energy in Nigeria, as compared to that in the US, is almost infinite. However, the environmental factors operating in our country are different from those in the United States. So, even if a major solar breakthrough were to be made in the US, the application in Nigeria would be very difficult unless reliable environmental data were compiled by local scientists and engineers. This is the same position with regards to the nuclear-power plants. We need basic research and development. The sooner we start, the better. International cooperation is essential. My concern is that we be in the position of technological readiness, so that, whether nuclear fusion or improved nuclear-fission power plants are developed by HICs, we are able to participate meaningfully in the breakthrough.

6. Environmental Concerns

The issue of the environment has been featured, world wide, in the last few decades, with special attention directed to nuclear energy. One of the first concerns was with fallout from atmospheric testing of the atomic bomb. The woıld outcry against atmospheric testing led to a general agreement by nuclear powers to ban atmospheric testing. As a result, testing is now done primarily underground. The only exception is France. With the emergence of the GREENS, even France is finding it difficult to conduct atmospheric tests of its nuclear bombs.

The major difficulty now, of course, is the nuclear-waste problem. I would like to comment on the issue of waste management, from the viewpoint of an African scientist. What concerns me, and I believe concerns people of like mind, is the amount of hypocrisy that goes on in the discussion of the problem of nuclear-waste storage arising from the operation of nuclear-power plants. A great deal of noise is made about the dangers posed by radioactive waste resulting from the operation of power plants. But hardly anything is said about the nuclear wastes arising from the production of nuclear armaments. It seems really a pity that African nations and other poor nations of the Third World are powerless to make this point, either through the international fora or in bilateral contacts. The position is worsened when you have the kind of dramatic and disturbing accidents, such as happened in Chernobyl. The radiation fallout and its effects on animals (milk) soon spread over a large portion of the world. For the first time, it dawned on African and other Third-World countries that, no matter what they do, they are part of one world. Situations like this demonstrate the concept of one earth, which has been quite a popular theme in WCC circles. For the first time, the United Nations organization appears to be fulfilling its role through its specialized agencies. Most people would agree that the role of the IAEA in the Chernobyl accident is something that everybody should be proud of, and that the organization should be congratulated. It further demonstrates what international collaboration in the area of peaceful uses of nuclear energy can achieve.

It is my view that it was primarily out of fear, rather than some noble concept, that any Third-World nations (together, of course, with the highly developed countries) signed the convention on early notification of nuclear-energy accidents. There were 72 signatories and 32 parties to this convention, out of 113 member states of the IAEA. At first reading, the signing of the convention on "Early Notification of Nuclear Accidents" seems a bit ridiculous in our African context. With the exception of the nuclear reactor in Zaire, I know of no other nuclear reactor in black Africa, excluding apartheid South Africa. Nevertheless, we need the convention, to protect us against accidents involving transportation of radioactive materials in the LDCs.

Another reason, which made a number of the LDCs more determined to be involved in international collaboration, on the problems of the environment, is the issue of dumping nuclear and toxic wastes. In Nigeria, we had a sad experience in 1988 when an Italian company, using agents in Nigeria and making negotiations with semi-illiterate citizens in the coastal port of KOKO, dumped hundreds of containers of dangerous, toxic chemical wastes in KOKO. Simple geiger-counter measurements showed that the radioactive levels in a number of containers were much greater than that permissible by international standards. All these events have added to the awareness of nations such as Nigeria and other African countries. Something has to be done to save and maintain the environment.

7. Nuclear Proliferation

From the historical background discussed earlier, the use of the atomic bomb in Japan demonstrated, very dramatically, that the possession of nuclear technology implied possession of awesome military firepower. Therefore, given the status of the scientific and technological infrastructure in Europe and the USSR, it was not conceivable that the US would be allowed the monopoly of this awesome military power. In 1963, the French carried out their A-bomb test in the Sahara Desert. It is interesting to observe that this test by the French, on African soil, produced radioactive fallout which was picked up by simple geiger counters in Nigeria. This test triggered angry protests leading almost to the severance of diplomatic relations with France. The only positive gain was the setting up of the Federal Radiation Protection Service (FRPS). This agency has been functioning till today.

There were essentially two developments in the acquisition of nuclear weapons. The first was that more and more countries soon joined the US as members of the elite nuclear club. This is horizontal proliferation. But there also was another direction in the development of nuclear weapons, that of the hydrogen bomb, with greater and greater killing capacity, as well as sophisticated delivery systems. Since we are not involved in any technical discussion on the manufacture of bombs, it is sufficient to know that one megaton H-bomb is equal to 50 Hiroshimas. At the time the WCC met in MIT in 1979, it was reported that both the US and USSR were making nuclear armaments at the rate of 10 megaton bombs per day, i.e., the capability for destroying 500 towns of the size of Hiroshima per day! The figures now may be considerably greater. The delivery systems have considerably improved, both in accuracy and in deception mechanisms. This development is referred to as "vertical proliferation."

Until recently, the concern of the Western and Eastern blocks - HICs - has been on horizontal proliferation. This was because it soon became apparent that more and more countries with the political will and the technological infrastructure can make their own nuclear bombs. Very little effort was made to criticize the vertical proliferation.

8. A Case Study: The Indian Experience

In order to give some perspective on the problem of nuclear proliferation as it concerns African countries in particular and the Third World in general, some description of the Indian experience is worthy of review. Soon after the Second World War, India obtained its independence from Britain. Pandit Nehru, the great statesman, became India's first Prime Minister. One of his earliest and major decisions was to invite the famous Prof. Bahba, eminent nuclear physicist, to return to India. He mandated Prof. Bahba to tour the entire world, to implore Indian scientists and engineers to come back home from various research institutes. A site was chosen in Trombay, near Bombay, for setting up India's first Atomic Research Centre. At first, the Indian scientists and engineers built a small reactor, on the principle of self-reliance, with the cooperation in the supply of uranium from Canada. Soon, many allied nuclear institutes, e.g., chemical processing, electronics, etc., were set up. Twenty-five years later, India exploded what it termed a "peaceful nuclear device." The world knew better. India had developed the capability of making the A-bomb. Recently, India tested a long-range rocket which I suppose is also a peaceful rocket! The strategy of India, i.e., the methodical way in which manpower was assembled, the political will demonstrated by the great statesman, Pandit Nehru; and the principle of self-reliance, coupled with some international collaboration, has been used by quite a number of Third-World countries to develop the capability to have their own bombs.

9. Non-Proliferation of Nuclear Weapons Treaty (NPT)

The purpose of the NPT can be traced to the original, Irish-Draft UN resolution in 1958, which drew the world's attention to the "danger of a possible spread of nuclear weapons." After years of negotiations, the Treaty on Non-

Proliferation of Nuclear Weapons (NPT) was opened for signature in Moscow, Washington, and London on July 1, 1968. A synopsis of the NPT as produced by the International Atomic Energy Agency (IAEA) is reproduced below.

> The main provisions of the Treaty are contained in Articles I, II, and III. In Article I the nuclear-weapon states undertake not to transfer nuclear weapons or other nuclear explosive devices or control over them to anyone whatsoever. This undertaking is spelled out in much detail. The nuclear-weapon states also undertake not to help any non-nuclear-weapon state to acquire nuclear explosive capacity in any way. Article II contains the reciprocal undertaking by non-nuclear-weapon states neither to acquire these weapons or devices nor even to seek or receive assistance to this effect.
>
> The verification of these obligations is ensured by the application of international safeguards through agreements to be concluded with the International Atomic Energy Agency (IAEA) (Article III). The rights of non-nuclear-weapon states to undertake research, production and exploitation of nuclear energy for peaceful purposes and to receive due assistance are reaffirmed (Article IV). Article V is designed to ensure that the benefits of peaceful nuclear explosions should be made available to all parties in accordance with the appropriate arrangements. Article VI contains the undertaking that the nuclear-weapon states will pursue further negotiations on disarmament in good faith.
>
> The right to conclude regional treaties to ensure the total absence of nuclear weapons is also reaffirmed in Article VII, and in Article VIII a review of the operation of the Treaty is foreseen five years after its coming into force.[2]

It is significant that Nigeria was a member of the 18-nation Disarmament Committee set up in 1965 to study the various draft proposals. It was one of the first group of countries who signed the NPT in 1968. Although Nigeria was hesitant in signing the safeguards agreement, it did so on February 25, 1988.

The problem with safeguards is that, right from the beginning, the nuclear "have nots" considered the treaty as one sided, despite all efforts to link it with nuclear disarmament. By 1957, when concern was shown, the US, USSR, and UK had fission and fusion bombs. Therefore anyone who wants to cheat on the treaty may feel morally able to do so. Also, on the technical side, there are many points on the nuclear fuel cycle where diversion can take place. The two most important are during enrichment and reprocessing. The enrichment facilities can be a major point of diversion. (The spent fuel contains only reduced amounts of uranium-235, in addition to significant amounts of plutonium-239.) There have been cases of hijacking of reactor fuel. But hijacking is very difficult, except when there is collusion between the exporting country and the hijackers.

The IAEA is well equipped technically to inspect facilities of countries that have submitted to the safeguards agreement. However, under the principle of national sovereignty, a country which had volunteered to accept safeguards may choose to open only some of its nuclear activities to be inspected. A country may claim, as happened during the Iran-Iraq war, that it is not possible for the IAEA inspectors to visit its country. During such intervals, diversion can take place.

For Third-World countries like Nigeria, there is a major advantage in having a safeguards agreement. Unless the Third-World country is associated with a neutral, multilateral organization such as the IAEA, it will be difficult for its nuclear project of any kind to take off, particularly at this point in time. Brazil and India have gotten a head start without the NPT. But even then Brazil has had to enter tougher agreements with West Germany before receiving the supply of the $2 billion US nuclear package.

The main disadvantage of the NPT is that, while it deals with horizontal proliferation, it does nothing about the vertical proliferation. Every day, the superpowers are improving the effectiveness of their nuclear arsenal and their delivery systems. Another disadvantage is that, because of the manifest inequity, countries who have the nuclear infrastructure, the resources, and the know-how, can opt out of the NPT.

10.Limitation of Nuclear Arms

As pointed out in the synopsis of the NPT, Article VI requires the nuclear-weapon states to pursue further negotiations on disarmament in good faith. "Treaties, protocols, international conventions, and unilateral restraints, while not true solutions, can buy time and move us in the right direction."[3]

The first major attempt at arms limitation is what is now known as SALT I. At the ceremonies for signing NPT on July 1, 1968, both Prime Minister Kosygin in Moscow and President Johnson in Washington agreed to undertake Strategic Arms Limitations Talks (SALT). After 127 sessions, the SALT Anti-Ballistic Missile (ABM) Treaty and Salt Interim Agreement came into force on October 3, 1972. Essentially the agreement was a "freeze" on the number of systems already deployed or in the process of being deployed. Specifically, two ABM sites were allocated to each country. 1710 intercontinental ballistic missiles (ICBMs) and submarine-launched ballistic missiles (SLBMs) in the US and 2358 in the USSR were allowed. Major problems arose, though not unforeseen during the negotiations, on verification. Both the treaty and interim agreement emphasize the so-called "national technical means" of verification. We know that the system of verification is a complex array of satellite monitoring, electronic listening posts in countries near the superpower being observed, aerial photographs, as well as naval listening posts. Despite all these means of verification, the tendency has been to overrate the armaments of the enemy and to underrate one's own arsenal, for obvious political reasons.

SALT II was a logical consequence of SALT I. It was embodied in SALT I which made provision for a review every five years. SALT II was more extensive. It set numerical limits on all the major types of strategic offensive systems, e.g., it included aircraft not mentioned in SALT I and set limits on the number of warheads each delivery vehicle can carry. (By this time, the superpowers had MIRVs, multiple, independently targetable re-entry vehicles.) Despite the fact that SALT II was signed in late 1979, the Senate of the US and former President Reagan blocked ratification as recently as 1988. Problems with SALT II were due largely to the fact that it made agreement on verification more difficult.

We end the discussion of this depressing topic on a note of optimism. In 1988, President Reagan and General Secretary Gorbachev signed and started to implement the Intermediate Range Nuclear Forces (INF) Treaty. It is a good beginning. There has been a remarkable thaw in the cold war. The two superpowers that started the arms race are cooling things down. There is hope. Already this reduction in the cold war is helping us in the African continent in Angola and Namibia. Let us hope that it will also lead to peaceful resolution of the South-African problem.

11. Conclusion

Two slogans of the WCC come to my mind in concluding this presentation. These are, "planet earth" and a "just, participatory and sustainable society" (JPSS). If all nations agree that we all, rich and poor, belong to the planet Earth, then the problems arising from peaceful uses of nuclear energy, such as nuclear-power plants, can be solved by cooperation.

Also, there would be no need for either horizontal or vertical nuclear proliferation, if the nations of the world attempt to build their societies on principles of JPSS. Then the positive issues to be resolved can be summarized under the following headings: justice, participation, and sustainability.

Notes

1 K. Tsipis, "Nuclear Power and Energy Needs of Third-World Economies," WCC background paper, Geneva, WCC, 1987.

2 "IAEA, A Short History of Non-Proliferation," Vienna, IAEA, February 1976.

3 H. York, "Arms Limitation Strategies," *Physics Today* (1983): 24.

Future Energy Choices in Light of the Greenhouse Effect

Ann Heidenreich
Environment Liaison Centre International
Nairobi, Kenya

A consensus is emerging in the world scientific community that global climatic change, caused primarily by current patterns of energy production and use, is the most urgent environmental issue facing the world community today.[1] The gravity of the problem of global climatic change is underlined by the great amount of activity at inter-governmental levels to understand this phenomenon, to predict its impacts on the environment and human populations, and to find solutions. Under the auspices of the Inter-Governmental Panel on Climatic Change (IPCC), scores of high-level scientific and policy meetings are taking place all over the world. Set up only recently (November 1988), the IPCC presented its interim report in August 1990.[2] This report will form the basis for negotiation of a Climate Convention, which some sources within UNEP suggest may be adopted as early as October 1991.[3] A Protocol restricting CO_2 emissions could be ready for signing by June 1992 at the United Nations Conference on Environment and Development. Such rapid conclusion of international agreements would be unprecedented.

1. Global Climatic Change

Atmospheric concentrations of the so-called greenhouse gases (ghg's) are increasing rapidly. The most important greenhouse gas, carbon dioxide (CO_2), has increased from pre-industrial levels of about 275 parts per million (ppm), by volume, circa 1750[4] to about 350 ppm estimated for 1988.[5] Greenhouse gases absorb infrared radiation (heat) rising from the earth's surface and re-radiate it back to earth, causing a warming effect. Carbon dioxide is responsible for about 50% of global warming, with the other major contributors being CFCs, N_2O, methane, and tropospheric ozone.[6] Energy production and use, which results in emissions of methane, N_2O, and ozone in addition to CO_2, is responsible for an estimated 57% of total greenhouse-gas emissions.[7]

The possibility that global warming could result from atmospheric pollution has been known for about 100 years.[8] Scientists are able to use crude mathematical models to simulate the chemical and physical processes which produce climatic change. The observed global average temperature rise over the past 100 years is consistent with the observed increases in greenhouse-gas emissions.[9] Half of the increases in ghg concentrations have come in the last 30 years,[10] and five of the hottest years on record occurred during the 1980s. Nevertheless, there are still scientific uncertainties regarding cause and effect, and substantial uncertainties regarding the ultimate impacts of global warming.[11]

It is known, however, that the impacts will be severe, causing serious changes in climatic patterns and a rise in sea level which will flood major population areas. Many scientists now agree that, despite uncertainties, there is sufficient cause for alarm to begin taking action now. The longer we wait, the more severe the damage will be.[12]

2. Global Warming in an Ecological Perspective

When I meet sceptics who don't believe global warming is an issue, I remind them that, on a global scale, we are dumping about 5.5 billion tonnes of carbon into the atmosphere every year from the burning of fossil fuels, and another estimated 1.8 billion tonnes of carbon from the destruction of tropical forests.[13] The earth is a closed system. Where do we think this carbon will go? There are natural carbon "sinks" in the oceans; trees and other vegetation also absorb CO_2 through photosynthesis, but these sinks can only absorb limited amounts of carbon. How can we have the audacity to think that the carbon and other wastes which we continually dump into our environment are simply going to disappear and not have any effect? In my opinion, it is simply a matter of common sense that dumping so much waste into the atmosphere will have an effect. I believe we are seeing that effect now in the climatic changes which are taking place.[14]

Environmental pollution has increased directly in proportion to the amount of waste we dump into the environment. As the wastes have increased, environmental pollution has gone from local (Love Canal) to regional (acid rain) to global. We have now discovered, to our horror, the ultimate pollution, pollution of our global atmosphere which threatens to destroy us all. Mother Earth, Gaia, the Goddess, is giving us a final warning. Judging from the amount of concern and frantic activity at international scientific and political levels, many people are scared. But unless the problem of climatic change is placed within a global environmental perspective, and solutions found within an ecological framework, I do not believe we will find our way out of this difficulty.

Generally, policymakers view energy issues from an economic perspective, the latest of these approaches being the so-called "least-cost" energy strategies based on an end-use analysis. By focusing on end uses rather than aggregate energy consumption and supplies, analysts have found that energy systems and technologies can be modified to supply the same services using less energy. And it is cheaper to save energy through the introduction of new technologies than to produce more energy.[15] These "least-cost" energy strategies do not take environmental externalities into account but prefer to make their point on conventional economic grounds.

However, I would like to suggest another entry point into the energy question, one that I think is more in line with the churches' idea of "integrity of creation." I think that this entry point can help us to evaluate the various energy options in the light of greenhouse warming and the other environmental issues we face. It has to do with what I shall call the "ecological principle."

3. The Ecological Principle

The "ecological principle," stated simply, says that we should use resources in a way which approximates as nearly as possible the way resources are used in natural ecological systems. This means a maximization of recycling so that resources can be reused by the system, and a minimization of waste. The moral imperative here is that we should use resources in such a way that they are available for future generations; we should *use* them, but we should not use them *up*.

The first time I heard this "ecological principle" stated was by Native American Mohawk Indians in upstate New York. They say you should use resources in such a way that they are available for seven generations. They and other native peoples believe in the integrity of creation and, in their traditional societies, use resources accordingly. They recycle.

To make one thing clear, when I talk about recycling, I don't mean the kind which takes industrial waste and turns it into a "useful" product, in order to get rid of the responsibility of disposing of it. (An appropriate example here would be plutonium "recycling" in fast-breeder reactors.) This type of "recycling" only puts off the problem of dumping waste, in an unusable form, back into the environment at a later date. When I say "recycling," I mean actually returning resources to the environment so that they resume their original state and can be reused by future generations.

It seems to me that there is a direct, inverse relationship between recycling and waste. The more you recycle, the less you waste. Furthermore, I believe that waste is the main cause of environmental degradation. In traditional societies, there is almost no waste. Traditional societies, free from the pressures caused by modern "development," return to the environment what they extract from it for their survival. They are sustainable because people do not upset natural ecological balances, but live in harmony with them. They do not *use up* resources.

Modern society, on the other hand, extracts resources from the environment (in the form of soil nutrients, timber, minerals, metals, fossil fuels, uranium, and so on) and spreads them to the four corners of the global market. It uses them up and throws them away, so that, for the most part, there is no way they can be returned to natural ecological systems for use by future generations. Not only does this deprive future generations of their birthright in the process, but it also irreversibly damages ecological systems by removing essential elements (causing, for example, soil degradation) and by clogging them with material that does not belong there (like air pollutants).

The conclusion we should probably draw from modern resource use is that, if you can't use a resource in such a way that it can be recycled, you shouldn't use it. Or, at most, you should use it only sparingly, because in using it, you will produce waste which will probably play havoc with natural ecological systems.

What does this have to do with energy? In order to get energy, we use resources. Some we just use and others we use up. Some, such as woodfuel, are in between. In general, those we use up also produce waste and disrupt natural ecological systems. Let us look at some of the various energy choices in the light of the "ecological principle."

4. Fossil Fuels

The main contributor to the anthropogenic greenhouse effect is carbon dioxide from the production and use of fossil fuels. Fossil fuels are energy resources which we use up, and in the process, we create serious environmental pollution by dumping the waste products back into the environment. While there are other wastes which result from the burning of fossil fuels (S_2O, NO_x, particulates, and so on), we shall focus on carbon waste here since this is the main contributor to global warming. As was mentioned above, more than 5.5 billion tonnes of carbon in the form of CO_2 are released annually to the atmosphere from the burning of fossil fuels. This carbon had been trapped in coal, oil, or gas for millions of years, and it cannot be reabsorbed back into natural ecological systems on any kind of human time scale, so it goes into the atmosphere, upsets atmospheric balances, and causes global warming.

It used to be that when we talked about the limits of growth, we meant that resources were finite and would eventually run out. Now, in the case of fossil fuel, it seems that the "limits imposed by the ability of the biosphere to absorb the environmental impacts of...energy consumption will be reached well before fossil fuels are exhausted."[16]

Some scholars have proposed that we compensate for CO_2 emissions from the energy sector through a major reforestation effort, creating "sinks" to absorb these additional emissions. However, to absorb all the carbon generated by fossil fuels would require the reforestation of an estimated area of about 300 million hectares a year, at a cost of $120 billion (US).[17] (It is interesting in this context to note that, while this sounds expensive, it is not much when compared with the nuclear alternative, which could run as high as $9 trillion (US).[18]) To date, forests are still being cut down about ten times faster than they are being replanted,[19] and it is clear that, while major reforestation efforts would help, it is more important to conserve the existing forests, and more important still to address the major source of CO_2 emissions, fossil fuels.

In much of the recent literature and debate on climatic change, one finds fairly general agreement that fossil fuels will have to be phased out.[20] The question is, how much and how fast. The most commonly mentioned target is 20-percent reduction (based on 1988 levels) by the year 2005. Twenty non-governmental organizations (NGOs) have called for 20- to 30-percent reductions by the year 2000 and 60-percent reductions by the year 2015.[21]

There are various aspects of the discussion about "how much, how fast." First of all, do we want to accept a certain degree of anthropogenic warming, say 0.1 degree Celsius per decade (as suggested by the Villach-Bellagio analysis[22]), or should we strive to stabilize the composition of the atmosphere and bring to zero the rate of anthropogenic warming (as many NGOs are demanding[23])? In a joint statement, the German Meteorological Society and the German Physical Society have concluded that, according to present knowledge, the global mean surface temperature should not rise by more than 1 degree Celsius above the current level, because even such a seemingly modest temperature increase could have noticeable impacts on the regional climate.[24]

After deciding what temperature increase is "acceptable," it is still necessary to figure out to what extent emissions will have to be controlled to achieve that target. Florentin Krause and others[25] have calculated that, in order to limit global warming to 0.1 degree Celsius per decade, global releases of carbon dioxide should have returned to present (ca. 1985) levels by about 2005; a 20-percent reduction below 1985 levels should be achieved by about 2015; a 50-percent reduction should be achieved by about 2030; and a 75-percent reduction should be achieved by about 2050. Since industrialized countries release much more CO_2, they will have to make relatively greater reductions of 20 percent by 2005: 50 percent by about 2015; and 75 percent by about 2030. The longer we delay the decision to reduce emissions, the greater will be the required reduction rate. Admittedly, some scientists are sceptical about such calculations because of insufficient understanding of biofeedback systems, including the sources and sinks of CO_2.[26]

In the meantime, reduction targets like the 20-percent one set at the World Conference on the Changing Atmosphere, in Toronto, are derived mainly on the basis of what is thought to be technically feasible, not on the basis of what is ecologically necessary. The question of "how much, how quickly" involves value judgments. How much are we willing to change the way we do things now? Some say the necessary changes are comparable to a new industrial revolution.

Once we have agreed to the phase-out of fossil fuels, the next question is, where will our energy come from? Many energy experts agree that the cornerstone of policies to reduce CO_2 emissions is energy conservation and efficiency. There is substantial potential for large energy savings in all of the areas where fossil fuels are used (industry, transport, electricity generation, military, households, and agriculture). We will briefly say something about these potentials, and then we will evaluate several alternative energy sources in terms of the "ecological principle." Since nuclear power is the main focus of this chapter and volume, we shall give more time to this energy source.

5. Energy Conservation

Improved efficiency of energy use is the key to reducing global CO_2 emissions. According to a group of international scientists who met under the auspices of the newly established Stockholm Environment Institute in August 1989, "energy conservation and efficiency measures, already widely available, could drastically reduce (often by as much as a third to a half) the rate of CO_2 build-up in the atmosphere."[27]

In their study, *Energy for a Sustainable World*,[28] Jose Goldemberg, *et al.*, compare two possible energy scenarios for the year 2020. One is a high, business-as-usual scenario which would approximately double fossil-fuel use and substantially increase nuclear power. The other is a low-energy scenario, including energy-efficiency measures (to hold energy use close to today's levels), reductions in coal and oil, and a very slight increase in nuclear. Both scenarios use a consistent set of assumptions for economic growth and levels of GNP during the period in question. Living standards in the North remain relatively unchanged, and those in the South attain European levels.

According to Amory Lovins,[29] the introduction of energy-efficient technologies, together with sustainable farming and forestry practices, would provide 85 percent abatement of global warming, while the phase-out of CFCs would displace the rest.

There are several ways to think about energy savings. The first is through a set of "technical fixes," such as more efficient automobiles, light bulbs, appliances, cookstoves, and electric motors. Replacing old, inefficient machines with new, efficient ones, will indeed go a long way to reducing greenhouse-gas emissions. A single example taken from Lovins should suffice to demonstrate the potential for these technologies.[30] According to this study, one quadrupled-efficiency, compact-fluorescent lamp, over its lifetime (which is about 13 times longer than for an ordinary incandescent lamp) saves hundreds of dollars investment in electrical supply, and avoids - assuming coal-fired electricity - about one tonne of CO_2 emissions and other pollutants. Such a lamp, displacing nuclear power, avoids the creation of a half-curie of strontium-90 and cesium-137 (main constituents of radioactive waste) and 25 mg of plutonium. Another recent study on the potential for energy savings from these new light bulbs found that they could save India and Brazil millions of dollars every day![31] Such "technical fixes" are essential in the short-term because of the urgent need to reduce CO_2 emissions. They are based on the assumption of high consumption societies which still *use up* the available resources.

However, in addition to reducing the energy intensity of particular activities, we can also switch to less energy-intensive activities. This second way of looking at energy savings involves making changes in the way we do things, so that the amount of energy required is reduced substantially, and all the resources involved are used in an ecologically sound, sustainable way. One example of this comes from the transport sector, where there is a real need to cut emissions. We can introduce more efficient automobiles, an essential, short-term step but, in addition, we shall have to reduce reliance on privately operated vehicles, use more bicycles or other means of non-motorized transport, improve and extend public transportation systems, and even redesign our urban centers to minimize the need for transportation, bringing living and working areas closer together. Farming practices will also have to change substantially, moving away from energy-intensive, fertilizer- and pesticide-dependent mono-cropping which is called "modern agriculture," to organic, low-energy-input agro-forestry and other mixed systems which more closely approximate natural ecological cycles. In addition to saving energy and conserving resources, such steps would have numerous other benefits as well, including improved health from reduced pollution, fewer road accidents, preventing erosion, and so on.

Underlining the importance of energy conservation as a crucial transitional measure to sustainable energy alternatives, Janos Pasztor says, in a paper prepared for the recent World Energy Conference in Montreal:

> It is recognized that even the most vigorous global program on alternative energy systems will not be able to substitute a sufficient fraction of present fossil fuel consumption to make a major impact on climate change. Similarly, even the most vigorous, and prohibitively costly nuclear program would only make a small impact on the climate change problem....Energy-efficiency measures, therefore, should be the cutting edge of all national energy policies during the transition to a more sustainable energy future.[32]

Improving energy efficiency will have a large, immediate, and cost-effective impact on carbon emissions. It costs less to save energy than to produce it and, since the energy saved does not have to be produced, improving efficiency will considerably reduce the environmental impacts of energy-generating systems.

So far, we have said that we will have to phase out fossil fuels. We have proposed that it is possible to replace most of them with energy conservation. But we will still need energy sources in the future. When the potential for energy savings is saturated, what then? The choices before us are nuclear power and renewable energy resources.

6. Nuclear Power

To begin with, let's subject nuclear power to the ecological principle. Like fossil fuels, nuclear power depends on a resource which gets used up. In the process, waste is produced in the form of radioactive materials which cannot be recycled in natural ecological systems and therefore do a great deal of damage (as explained in the paper of Rosalie Bertell in this volume). Nuclear power clearly violates the integrity of creation and, as such, should be written off.

I guess this amounts to "excluding the human use of nuclear energy on the basis of an *a priori* ethical principle," as referred to in "The Churches and the Debate about Nuclear Energy," a document distributed as background reading for the 1989 WCC Kinshasa conference.[33] If that is the case, so be it. However, it is probably a good idea to mention here other arguments concerning the nuclear-power option in relation to climatic change.

In "Faith and Science in an Unjust World," another background document distributed for the Kinshasa conference, it is stated that

> nuclear power, as such, cannot have a significant impact on reducing carbon-dioxide build-up for a very long time. On the other hand, conservation measures which would double the efficiency of automobile engines and/or

> building insulation could dramatically reduce the rate of CO_2 accumulation by the turn of the century. *Thus the CO_2 problem cannot be a decisive factor in deciding one way or the other on nuclear power for the short and medium terms.*[34] (my emphasis)

This argument has recently been developed in a study by Bill Keepin and Gregory Kats, entitled "Greenhouse Warming: Comparative Analysis of Nuclear and Efficiency Abatement Strategies."[35] Keepin and Kats analyze two popular strategies for reducing CO_2 emissions from fossil fuels: revitalizing nuclear power and improving energy efficiency. As a result of their study, they conclude:

> Under highly favourable assumptions for nuclear power in scenarios of moderate to high energy growth, we find that even if large nuclear plants (1000 MW) could be built every one to three days from now until 2025 (which is impossible in the Third World), global CO_2 emissions would still continue to grow. Thus, nuclear power cannot contribute significantly to abating greenhouse warming, except possibly in scenarios of low energy growth for which the problem is already largely ameliorated by efficiency improvement.[36]

Furthermore, say Keepin and Katz:

> In the USA - the world's largest producer of CO_2 - each dollar invested in electric efficiency displaces nearly seven times as much CO_2 as a dollar invested in nuclear power. Even if the most optimistic aspirations for the future economics of nuclear power were realized today, efficiency would still displace between 2.5 and 10 times more CO_2 per unit investment.[37]

By diverting funds from the more effective option, investment in nuclear power would actually exacerbate the problem of global warming, leading to more CO_2 emissions rather than less.

Keepin and Kats conclude that revitalizing nuclear power "would be a relatively expensive and ineffective response to greenhouse warming, and that the key to reducing future CO_2 emissions is to improve the energy efficiency of the global economy."[38]

Janos Pasztor concludes, in a paper circulated at the WCC Kinshasa conference, that

> the Climate Change problem, as important, and as serious it may turn out to be, does not justify a considerably increased global nuclear program. Even if for other political or other socio-economic reasons such an intensive global nuclear program were initiated, its impact on carbon- dioxide emissions would be only marginal.[39]

There are other arguments regarding the nuclear solution. Herman Damveld and Ian Tellam, in a paper written for the World Information Service on Energy[40], calculate that, if nuclear power would be expanded rapidly to supply

70% of global electricity, then uranium resources would be exhausted by 2011. They add that "current experience of fast-breeder technology indicates that fast-breeder reactors are not able to extend the lifetime of nuclear power."[41] The new fast-breeder reactor designed in Europe (EFR), they explain, will take 292 years of normal operation to produce enough plutonium to make a new core. Tellam and Damveld also point out that a certain amount of energy is used to mine uranium and, if fossil fuels are used for this purpose, a point will be reached, as uranium grades fall, where nuclear emits as much CO_2 as coal. "Therefore," they conclude, "a large expansion of nuclear power could, at best, make only a short and limited contribution to solving the greenhouse problem."[42]

A paper on climatic change produced by the Australian Conservation Foundation argues that

> building up a nuclear power capacity would commit us to an electric power system that would block a transition to solar power. Using solar to replace fossil fuels requires significant decentralization. Nuclear power, however, produces electricity in highly centralized plants. Cutting off our options with regard to solar power at this stage would be extremely short-sighted.[43]

But if we dispense with fossil fuels and exclude nuclear power as an environmentally damaging and unsustainable energy source, what energy technologies are left which can be used in an environmentally sound and sustainable way? The only energy sources free from significant climatic impact are the dispersed, renewable energy alternatives.

7. New and Renewable Sources of Energy

Biomass, solar, wind, small-scale hydro power, hydrogen, geothermal, tidal, wave, and ocean-thermal energy conversion all present attractive possibilities for the future. Each of these options has to be evaluated with respect to its environmental impacts. Since most of the renewable energy sources are site-specific, potential uses can only be evaluated within local or national contexts. Most of these sources are dispersed and relatively small, so they will be used in combination with others. Although a great deal of work is going on to develop all of these sources, the work is decentralized in nature. Thus it has been difficult to provide a good overview of the potential for these new and renewable sources of energy. We will look briefly at the more important of these new sources.

8. Biomass

Biomass as an energy source is used mainly for woodfuel, charcoal, biogas, methanol, and producers' gas. It is the most important source of energy for developing countries (76 percent of the world's population), since it provides 38 percent of their energy, equivalent to 21 million barrels of oil a day.[44] Deforestation is often attributed to the use of wood for energy, but recent studies have shown that clearing land for agriculture and commercial pressures is responsible for more deforestation than energy use. Biomass is, and will remain, a major energy source into the future.

But what about climate change? Does the use of wood and other biomass sources for energy increase the atmospheric concentrations of CO_2? The answer is "yes, if an equivalent amount of trees and other vegetation is not regrown to reabsorb the CO_2 emitted through burning." In Kenya, a study is now underway into the sources and sinks. It has tentatively concluded, with certain assumptions, that Kenya is a net CO_2 emitter with fossil fuel-based transport and woodfuel being the major culprits.[45]

However, assuming sufficient replanting is undertaken, all the CO_2 emitted through biomass burning will be reabsorbed again through photosynthesis. According to the UNEP study cited earlier, the "adverse impacts of biomass use may be avoided through sustainable management practices and use of improved technologies."[46] This is, however, not happening now, and substantial policy changes will be needed to achieve this goal.

9. Solar Energy

For countries with a lot of sun, solar power may be the most promising source of energy after conservation. An official government study in the US has shown that solar resources could be used to supply 40 percent of total energy needs in that country.[47] Energy from the sun can be harnessed in many forms, including solar water heating, passive solar heating (and cooling) of buildings, and photovoltaic panels.

The cost of photovoltaic cells has dropped substantially in the past 20 years, from $120 (US) per peak watt in the early 1970s, to $5 (US) at present.[48] Further price reductions and improvements in the efficiency of solar cells are expected. It has been argued, by Jose Goldemberg and his colleagues[49], that the new amorphous silicon cells have an energy density over the lifetime of the system of approximately 1200 kilowatt hours per gramme of silicon, almost five times the energy density of enriched uranium used in a conventional nuclear reactor (260 kWh/g). Goldemberg points out that the world has a lot more sand than uranium. He also notes that the nucleus of a uranium atom is available only once, and a silicon cell's "stored electronic energy can be extracted repeatedly with continual solar recharging" over its lifetime.

But while our energy future is definitely a solar future, we must also take care that our solar energy systems do not violate the ecological principle. It was reported to me that, at the recent World Energy Conference in Montreal, the only paper presented on solar energy dealt with solar satellites. One obvious question, when we talk about solar systems of this scale, is: "how much fossil fuel, and how many nonrenewable resources will be used up in the manufacture and deployment of all this equipment?" These questions have been eloquently raised by Jeremy Rifkin in his book: *Entropy, Into the Greenhouse World*.[50] According to Rifkin, truly huge amounts of nonrenewable resources would be necessary to build a massive solar infrastructure. Such massive systems, however, are not necessary and, on a smaller scale, solar systems can be produced in a sustainable way, so that resources are used, but not used up.

10. Small-Scale Hydro

Hydropower for mechanical uses and electricity generation is an established energy resource with considerable potential for expansion. Environmentalists tend to restrict the use of the term "renewable" to small-scale hydropower, because of the enormous amount of environmental damage and human suffering caused by many large-scale hydrodams. Since much of the damage done by large-scale hydro is irreversible, and because their lifetimes are relatively short, most of them could hardly be considered sustainable energy sources. Small-scale hydropower units, however, can function within ecological systems without causing serious disturbances.[51]

11. Conclusions

To solve the problem of climatic change, which threatens life as we know it on this planet, we have to change profoundly the way we use our resources. We have to learn how to use resources and return them to their natural ecological systems, so that those systems remain intact, so that the resources are available for future generations, and so that they are not turned into mountains of useless waste that foul our environment.

Fossil fuels are an unsustainable resource. They can only be used once, and they produce carbon waste which is largely responsible for greenhouse warming. A rapid reduction in the use of fossil fuels is the only way to bring about a substantial long-term reduction in CO_2 emissions.

The fastest and cheapest way to begin phasing out fossil fuels is by improving energy efficiency. Energy efficiency measures, comprising both new technologies and new lifestyles, will provide the transition to a more sustainable, renewable energy future fuelled by alternative, renewable energy resources.

Nuclear power is among the alternatives being promoted, but it is not a renewable, sustainable energy source. Moreover, it has been convincingly argued that nuclear power could not contribute substantially to alleviating the global greenhouse problem, and by diverting funds from more effective measures, it would actually make matters worse. That leaves us with the new and renewable energy resources, which are the real energy resources of the future. We have an urgent obligation to develop these resources in such a way that they comply with the "ecological principle," providing energy without using up resources or polluting the environment.

The implementation of both strategies, the efficient use of energy, and the development of renewable energy resources, will require the cooperation and commitment of all sectors of society. At the grassroots level, people will have to become aware of the relationship between their behavior and global and local environmental problems. Especially in the North, people will have to recognize that a barrel of oil used today, for whatever serious or frivolous purpose, is gone for good and, in its place, we have more of the air pollutants which are now threatening the life of our planet. This process of awareness-raising can be facilitated by local studies, carried out by church groups or other NGOs. Such studies can examine how human activities affect the environment and what people can do to stop environmental destruction. Because people do not usually take action to correct a problem until they see how it relates to them, such studies are a precondition for change. Local people can best design systems for reducing energy use and for promoting other forms of environmental protection which are appropriate for them. They will also play a major role in adapting renewable energy technologies to suit their specific needs and locally available resources.

National governments will have to incorporate environmental considerations into energy planning. A strong case is being made that the transition to a sustainable energy future cannot be left entirely to the market, and that various mechanisms will have to be found to internalize environmental costs in energy pricing.[52] This includes efforts to remove current subsidies from the most damaging forms of energy production and use and to encourage energy savings and the transition to renewables, through direct subsidies or other methods. Intervention in the market could be justified (if such a justification is needed) in this context because the objective is to preserve and/or improve the quality of an essential, free, public good (air), an objective which individual producers and consumers by themselves cannot achieve. It is the environmental equivalent of military defense.[53]

We began this chapter with a reference to the great deal of activity going on at the international level, and we have come full circle, back to the international level. The actors here are the intergovernmental agencies; the World Bank and other international funding agencies; multinational corporations; and international NGOs like the World Council of Churches and my own organization, the Environment Liaison Centre International. Intergovernmental agencies are moving rapidly towards a global climate convention. There are strong pressures on the World Bank to incorporate environmental considerations into its funding policies. Corporations are moving on the issue of CFCs, but they need to be pressured to introduce long-term moral and environmental considerations, in addition to short-term economic ones, into their policies.

International NGOs have a crucial role to play - by raising these issues with their constituencies, informing them, supporting local studies which link the global issues with their local situations, and stimulating debate on policy options. International NGOs can help create the broad public awareness and participation which is essential if sustainable solutions are to become a reality.

Notes

1 John M. Christensen and Myles R. Allen, "Climate Change and the Need for a New Energy Agenda," *Energy Policy*, to be published. See also Bert Bolin, *et al.* (editors), *Scope 29: The Greenhouse Effect, Climatic Change and Ecosystems*, Chichester, John Wiley & Sons, 1986, p. xxxi; hereafter cited as: Bolin.

"The possible problem of a change in climate due to the emissions of greenhouse gases should be considered as one of today's most important long-term environmental problems."

2 For information on IPCC activities, contact IPCC Secretariat, c/o World Meteorological Organization, P.O. Box 2300, CH-1211 Geneva 2, Switzerland.

3 Personal communication.

4 Bolin, p. xxvi.

5 *Policy Options for Stabilizing Global Climate*, Draft Report to Congress, Executive Summary, Washington, D.C., United States Environmental Protection Agency, Office of Policy, Planning and Evaluation, February 1989, p. 13; hereafter cited as: EPA Report.

6 EPA Report p. 12.

7 EPA Report p. 55.

8 Bolin, p. 2.

9 J. H. Karas and P. M. Kelly, *The Heat Trap*, London, Friends of the Earth, 1988, p. 4; hereafter cited as: Karas.

10 Bolin, p. xxv.

11 EPA Report, p. 83.

12 EPA Report p. 85.

13 Karas, p. 10, and David Hall and F. Rosillo-Calle, "CO_2 Cycling by Biomass: Global Bioproductivity and Problems of Deforestation," paper presented at the Conference: Amazonia: Facts, Problems and Solutions, July 31-Aug. 2, 1989, University of Sao Paulo, Sao Paulo, Brazil, p. 8; hereafter cited as: Hall. There is a fair amount of agreement on the figure of 5.5 billion tonnes of emissions for fossil fuels. Karas and Kelly say deforestation and changing land use are currently responsible for a net annual release of between 0.6 and 2.6 billion tonnes of carbon, but take 1.8 billion tonnes C/yr as the most likely estimate. Hall and Rosillo-Calle estimate that net emissions from tropical forest clearing in 1980 resulted in release of 1.8 billion tonnes C/yr. to the atmosphere as CO_2, but say that the current value is likely to be much higher.

14 Dr. Eldon Franz of Washington State University, currently Visiting Fellow at the International Council for Research in Agroforestry (ICRAF) in Nairobi, has the following to say regarding the argument from common sense: "I doubt that common sense alone would link the discharge of combustion products into the atmosphere to climatic change but there is abundant scientific evidence to support the common sense-seeming notion that emitting vast quantities of potentially reactive substances into the air will do something. Indeed, it results in large scale transformations of atmospheric chemistry with impacts affecting both terrestrial and aquatic ecosystems." Franz cites P. J. Crutzen and T. E. Graedel, "The Role of Atmospheric Chemistry in Environment-Development Interactions," in W. C. Clark and R. E. Munn (eds.), *Sustainable Development of the Biosphere*, Cambridge, Cambridge University Press, 1986, and T. E. Graedel and P. J. Crutzen, "The Changing

Atmosphere," *Scientific American*, September, 1989, pp. 58-68. (Personal communication)

15 See, for example, A. Lovins, *Least-Cost Energy, Solving the* CO_2 *Problem*, Snowmass, Colorado, Rocky Mountain Institute, 1989; hereafter cited as: Lovins.

16 J. M. Christensen, *et al.*, "The New Energy Agenda: The Role of New & Renewable Sources of Energy," First Draft, Prepared for the ACC Interagency Group on NRSE, Nairobi, UNEP Energy Unit, 1989, p. 5.; hereafter cited as Christensen, NRSE.

17 N. Meyers, "The Greenhouse Effect: A Tropical Forestry Response," *Biomass* 18 (1989): 73-78; cited in Hall, p. 27.

18 Hall, p. 27.

19 EPA Report, p. 71: "It is generally estimated that approximately 11 million hectares (Mha) of tropical forests are currently lost each year, while only 1.1 Mha are reforested per year." The EPA Report goes on to say, however, that "recent analysis of remote sensing data from Brazil ... suggest that in 1987, 8 Mha were cleared in the Brazilian Amazon alone."

20 Statement issued by the participants at the World Conference on the Changing Atmosphere--Implications for Global Security, 27-30 June 1988, Toronto, Canada, paragraph 22.

21 NGO Statement of Policy Options to Curb Climatic Change, Prepared for the UNEP Governing Council Meeting, 15-26 May 1989, Nairobi, Environment Liaison Centre International, p. 5; hereafter cited as: NGO Statement.

22 Jill Jaeger, "Developing Policies for Responding to Climatic Change," A summary of the discussions and recommendations of the workshops held in Villach and Bellagio, WHO/UNEP, 1988, p. 25.

23 NGO Statement, p. 4.

24 W. Bach, "Modelling the Climatic Effects of Trace Gases: Reduction Strategy and Options for a Low Risk Policy", Prepared for the World Congress: Climate and Development, Climatic Change and Variability and the Resulting Social, Economic and Technological Implications, Hamburg, West Germany, November 1988.

25 Florentine Krause, Wilfrid Bach, and Jon Koomey, *Energy Policy in the Greenhouse*, Volume One, El Cerrito, California, International Project for Sustainable Energy Paths, 1989. Executive Summary.

26 According to Dr. Eldon Franz (see endnote 14 above), recent studies suggest that the productivity of tropical grass ecosystems has been greatly underestimated by standard techniques calling all previous calculations of global carbon balance into question (S. P. Long and others, "Primary Productivity of Natural Grass Ecosystems of the Tropics: A Reappraisal," *Plant and Soil*, No. 115 (1989): 155-166). Feedback processes are controversial because they are just beginning to be effectively investigated and could either serve to amplify (A. Raval and V. Ramanathan, "Observational Determination of the Greenhouse Effect," *Nature*, No. 342 (1989): 758-761) or to damp (J. Mitchell and

others, "CO_2 and Climate: A Missing Feedback?" *Nature*, No. 341 (1989): 132-134) the climatic response to a greenhouse forcing. (Personal communication)

[27] "Implementation Measures for Responding to Climatic Change," Draft report of a workshop held at the Stockholm Environment Institute, August 1989, p. 9; hereafter cited as: Stockholm Environment Institute.

[28] Jose Goldemberg, Thomas B. Johansson, Amulya K. N. Reddy, and Robert H. Williams, *Energy for a Sustainable World*, Washington, D.C., World Resources Institute, 1987.

[29] Lovins.

[30] Lovins.

[31] A. Gadgil and G. De Martino Jannuzzi, "Conservation Potential of Compact Fluorescent Lamps in India and Brazil," Lawrence Berkeley Laboratory, Berkeley CA and Engineering Faculty, Universidade Estadual de Campinas, Sao Paulo, Brazil.

[32] J. Pasztor, "Toward Sustainable Energy Futures," Nairobi, UNEP, 1988, pp. 12-13.

[33] "The Churches and the Debate about Nuclear Energy," in *Faith, Science and the Future*, Preparatory Readings for the 1979 Conference of the World Council of Churches, Geneva, WCC, p. 115.

[34] "Faith and Science in an Unjust World," in *Faith and Science in an Unjust World*, Report of the World Council of Churches' Conference on Faith, Science and the Future, Geneva, WCC, Volume Two, p. 97.

[35] Bill Keepin and Gregory Kats, "Greenhouse Warming: Comparative Analysis of Nuclear and Efficiency Abatement Strategies," in *Energy Policy* 16, No. 6 (December 1988): 538-561; hereafter cited as: Keepin and Kats.

[36] Keepin and Kats, p. 538.

[37] Keepin and Kats, p. 538.

[38] Keepin and Kats, p. 538.

[39] J. Pasztor, "What is the Role of Nuclear Power in Slowing Down Global Warming?" Draft, Boston, Stockholm Environment Institute, 1989, p. 39.

[40] H. Damveld and I. Tellam, "No Place for Nuclear Power in the European Greenhouse," Amsterdam, WISE, 1989; hereafter cited as: Damveld and Tellam.

[41] Damveld and Tellam, p. 1.

[42] Damveld and Tellam, p. 1.

[43] W. L. Hare and S. Prior, "The Contribution that Australian Industry, Science and Technology Can Make in Reducing the Greenhouse Effect," Submission to Inquiry by the Senate Standing Committee on Industry, Science and Technology, Australian Conservation Foundation, September 1989, p. 60; hereafter cited as: Hare.

[44] Stockholm Environment Institute, p. 12.

[45] Dr. Herick Othieno and Dr. Eldon Franz, "Energy Contribution to Climate Change: The Kenyan Role," Nairobi, 1989, unpublished paper.

[46] Christensen, NRSE, p. 15.

[47] L. S. Johns, *et al.*, *Application of Solar Technology to Today's Energy Needs*, Washington, D.C., Office of Technology Assessment, 1978, Vol. 1, p. 3.

[48] Joan M. Ogden and Robert H. Williams, *Solar Hydrogen, Moving Beyond Fossil Fuels*, Washington, D.C., World Resources Institute, 1989, p. 17.

[49] Cited in Hare, p. 42.

[50] Jeremy Rifkin, *Entropy, Into the Greenhouse World*, New York, Bantam Books, 1989.

[51] On the question of how much damage constitutes a serious disturbance, Dr. Eldon Franz (see endnote 14 above) says that disturbances can only be judged to be serious or not in relation to the nature of "impacts" or state changes which result from them (Odum, *et al.*, "Perturbation Theory and the Subsidy-Stress Gradient," *Bioscience* 29, No. 6 (1979): 349-352). Such state changes include any readily detectable population responses to disturbance (F. A. Bazzaz, "Characteristics of Populations in Relation to Disturbance in Natural and Man-Modified Ecosystems," in H. A. Mooney and M. Godron (editors), *Disturbance and Ecosystems: Components of Response*, New York, Springer-Verlag, 1983, pp. 259-275). To be detectable, the change due to disturbance (the signal) must be large relative to the intrinsic variability (the noise) in the system. Franz and Bazzaz ("Simulation of Vegetation Response to Modified Hydrologic Regimes: A Probabilistic Model Based on Niche Differentiation in a Floodplain Forest," *Ecology* 58, No. 1 (1977): 58-68) have demonstrated that population or community responses can be used to estimate impacts of large hydro projects and to determine which designs will produce detectable changes and which designs will not. Serious disturbances can therefore be avoided by using ecological assessments to guide their design. The same principles can be applied to small scale hydro as well. (Personal communication)

[52] See, for example, Christensen, NRSE, p. 20-21.

[53] Stockholm Environment Institute, p. 19.

What Role Can Nuclear Power Play in Mitigating Global Warming?

Janos Pasztor
Stockholm Environment Institute
Boston, Massachusetts

1. Background

> Humanity is conducting an unintended, uncontrolled, globally pervasive experiment whose ultimate consequences could be second only to global nuclear war. The Earth's atmosphere is being changed at an unprecedented rate by pollutants resulting from human activities, inefficient and wasteful...fossil fuel use and the effects of rapid population growth in many regions. These changes represent a major threat to international security and are already having harmful consequences over many parts of the world.
>
> Far-reaching impacts will be caused by global warming and sea-level rise, which are becoming increasingly evident as a continued growth in atmospheric concentrations of carbon dioxide and other greenhouse gases. Other major impacts are occurring from ozone-layer depletion resulting in increased damage from ultra-violet radiation. The best predictions available indicate potentially severe economic and social dislocations for present and future generations, which will worsen international tensions and increase risk of conflicts between and within nations. It is imperative to act now.

This statement is from the 1988 Toronto Conference on the Changing Atmosphere, one of a series of important international gatherings which have drawn attention recently to global warming.[1] This chapter addresses carbon-dioxide emissions only, the principal energy-related greenhouse gas,[2] responsible for about half of global warming.

In 1985 the total global anthropogenic emission of carbon dioxide was about 5 gigatonnes (Gt) of carbon, with approximately 25% from North America, 15% from Western Europe, 26% from Eastern Europe, 11% from China, 5% from Japan, and 18% from the rest of the world combined.[3] Long-range projections of global emissions vary considerably, depending on the particular assumptions. One scenario shows a slight decrease to 4.8 Gt by 2020,[4] while another shows a substantial increase to 12 Gt,[5] climbing sharply to 26 Gt by 2050. More extreme projections have been made on the low and high sides.[6] For the purposes of this study, we consider 5 Gt as the lower and 20 Gt as the upper bounds of yearly, global, carbon-dioxide emissions by 2020.[7]

The Toronto Conference has called for an initial step of 20% reduction in annual greenhouse-gas emissions by the year 2005 (about 1 - 3 Gt/yr) through various measures, including energy efficiency and alternative energy supplies.[8] Although technically feasible, such a target is by no means easy to achieve. A number of options are available, however, on both the demand and supply sides.[9] The options differ from country to country, and have different ranking, depending on local economic, political and social contexts. This menu includes efficiency measures everywhere. The main supply options are nuclear energy, advanced fossil conversion systems, as well as solar and biomass energy. Nuclear fusion and other futuristic energy systems such as the hydrogen economy are not expected to play any important role during the coming three decades, the period considered in this study.[10]

The nuclear fuel cycle, during normal operations, emits little carbon dioxide. Upstream carbon-dioxide emissions vary with the source of electricity used in the preparation and fabrication of the nuclear fuel. According to US data, the BWR fuel cycle emits approximately 90 times less carbon dioxide than a conventional coal-based power plant.[11] According to data from the Federal Republic of Germany, reflecting different fuel cycle details, the PWR fuel cycle emits 20 times less carbon dioxide than the coal fuel cycle.[12] Consequently, nuclear energy is a low carbon-dioxide emitter, and can, in principle, be considered as an alternative source of energy in greenhouse-gas-reduction strategies. As an added bonus, the nuclear fuel cycle emits only small amounts of acidifying gases and other air pollution typically caused by the combustion of fossil fuels.

The purpose of this chapter is to look at nuclear fission, as one of the available options, and evaluate the extent to which it can play a role in greenhouse-gas-reduction strategies during the next three decades, covering the short (approximately the first decade), and medium terms. The study will not develop detailed, quantitative scenarios of expected nuclear developments, but rather evaluate qualitatively the extent to which nuclear energy may develop during the period in question. No attempt will be made here to make a quantitative comparison with alternative energy supplies, or with energy-efficiency measures.

2. Nuclear Energy in the 1980s

At the end of 1988 there were a total of 429 commercial nuclear reactors in operation, with a capacity of 311 GWe. In addition, there are 105 reactors under construction, representing an additional capacity of 85 GWe.[13] During 1988, about 17% of the world's electricity was generated from nuclear energy. The total world operating experience with nuclear reactors passed the 5000 reactor years mark at the end of 1988.[14] These figures indicate that nuclear power has entrenched itself as a major source of electrical energy.

A more detailed analysis of these figures, however, shows a different (and less optimistic) situation. There has been a considerable slow-down in the ordering of, and planning for, new reactors. Many old projects have been canceled. In

1988 construction started on only six new plants, with 14 planned to begin in 1989.[15] As a result, apart from a handful of countries, such as France and Japan, where the nuclear industry is doing relatively well, the immediate future of nuclear energy does not look very promising.

The International Atomic Energy Agency (IAEA) estimates that total world nuclear capacity will rise to 490 GWe, of which 320 GWe will be in OECD countries, 120 GWe in the centrally planned economies and 50 GWe in the developing countries. The World Energy Conference (WEC) also scaled back its projections of future nuclear energy use. In 1983 the WEC projected that 12% of global energy supplies would be provided by nuclear energy in 2020. At its 1989 congress in Montreal, this was scaled down to 7-8% which, nevertheless, would require a tripling of present nuclear capacity.[16]

2.1 The OECD Countries

Out of the 24 OECD countries, 12 use nuclear power. Those without nuclear power will all continue that way until 2005, according to present plans, except for Italy, which expects a 1.2 GWe capacity on line by 2005. The total installed nuclear capacity for OECD countries was 247 GWe in 1988; this is expected to rise to 313 GWe by 2005. Actual generation of nuclear electricity represented 24% in 1988, and should drop to 22% in 2005.[17]

Japan is the only country where significant expansion in nuclear generating capacity is expected by the year 2005. In addition to plants already operating or under construction, only four new reactors in OECD countries are firmly committed, and all those are in Japan. There are only 17 plants which are in a planning status in the whole of the OECD, of which 14 are for Japan, two for France and one for the UK, as of the end of 1988.[18] Following its policy of phased withdrawal from nuclear energy, Sweden will start to cut nuclear capacity from 9.7 GWe in 1988 to 4 GWe in 2005,[19] in line with its decision following a 1980 popular referendum, to phase out nuclear energy completely by 2010.[20]

Meanwhile, increasing numbers of existing nuclear reactors will come to the end of their lives. By 2005, 42 nuclear reactors with a total capacity of 13 GWe will be retired.[21] Given the present hesitance to plan and order new reactors, many may not be replaced by other nuclear units.

2.2 Centrally Planned Countries in Europe

Nuclear energy in the USSR and other centrally planned economies (CPE) in Europe has been an important cornerstone of development efforts. At the end of 1988, the total installed nuclear capacity was 43 GWe, of which 34 GWe was in the USSR. While a number of countries, such as the German Democratic Republic, Czechoslovakia, Poland, and Hungary, have access to cheap lignite for power production, environmental considerations have contributed to plans for increased nuclear-generated electricity and district heating. An additional 26 reactors (21 GWe) are under construction in the USSR, and 23 (15 GWe) in the other CPE countries.[22]

2.3 Developing Countries

The situation in the developing nations has, to some extent, paralleled that in the industrialized countries. In the early 1970s ambitious nuclear plans were developed, and a large number of orders were placed. India was to reach 10 GWe capacity by the year 2000, Mexico 22 GWe, Brazil 10 GWe and the Republic of Korea had plans for 46 reactors by the year 2000.[23] China did not launch a major near-term nuclear capacity, although expectations of 300 GWe and 250 GWt for heat by 2030 have been reported.[24]

The reality has turned out very different. At the end of 1988, installed nuclear capacity in seven developing countries was 15 GWe. This included the Republic of Korea (6.3 GWe), Taiwan (5 GWe), India (1.2 GWe), Argentina (0.9 GWe), Yugoslavia (0.6 GWe), Brazil (0.6 GWe), and Pakistan (0.1 GWe). In addition to this, 11 GWe of nuclear capacity are now under construction, including 0.7 GWe in Argentina, 1.2 GWe in Brazil, 2.1 GWe in China, 0.8 GWe in Cuba, 1.8 GWe in India, 2.4 GWe in Iran, 0.9 GWe in Korea, and 1.3 GWe in Mexico, making a total of about 26 GWe of nuclear capacity in operation or under construction.[25]

Argentina and India wanted to develop nuclear technology with local resources. In both cases, the possible military use of the nuclear science and the relevant technologies probably played an important part.[26] While in both countries this objective was partly met, neither is fully a success story. For example, India builds its own reactors, although at a heavy price for the nation as a whole. Furthermore, Indian reactors are beset with low availability factors due to a host of technical problems.

Both the Republic of Korea and Taiwan have well developed nuclear sectors, matching well the situation of two rapidly industrializing countries. Korea in the 1980s is largely a success story, although new construction and ordering has slowed down considerably. In addition, the original goal of energy independence has turned into a total dependence on foreign fuel- cycle services, and a continuation of the "special relationship" with the USA. Taiwan has accumulated 44 successful reactor-years of experience, with its reactors now generating 41% of its electricity supply. Yet, it has run into similar problems as industrialized countries with successful nuclear programs, including an overcapacity in electrical generation. Taiwan is now constructing 5 GWe of coal-fired facilities to generate electricity from cheap, imported coal in the 1990s.

2.4 Discussion

In the USA, all nuclear plants ordered since 1974 have been canceled, and none will be ordered in the foreseeable future. Capacity expansion will be delayed as long as possible, and when unavoidable, will be met from various advanced coal- and gas-fired schemes. Meanwhile, the same utility executives, who have decided not to order new nuclear plants, are convinced that nuclear energy does have the potential to be the cheapest, and environmentally most benign, source of electricity.[27] However, for this potential to be realized, major changes are required both in the nuclear industry and in its regulation.

While in a handful of countries nuclear appears to be doing well (e.g., Japan, France, Korea, and in some of the CPE countries), in most there has been a considerable slow-down (e.g., Germany, Switzerland, India), or outright cessation of nuclear developments (e.g., Sweden, USA, Mexico). A number of factors are contributing to the failure of nuclear power to penetrate the energy sector as planned. Depending on the country, these include unfavorable economics, heavy financial burdens during construction, concerns about radioactive waste disposal, and especially public perception of the high accident and public-health risk. Although the problem of nuclear proliferation is less intensely debated now than in the 1970s, it will remain an issue as more inter- and intra-national nuclear shipments are made in the coming decades.

3. New Technologies

There are a number of new technologies which may have important impacts on the penetration of nuclear power in the future. Some of them are evolutionary changes to existing technologies, while others include radically new designs on what has been called the "Second Nuclear Era."[28] Some involve only the reactors, while others have impacts at various stages of the fuel cycles. The main trends of technological change are summarized below.[29]

3.1 *Advanced Light-Water Reactors*

The nuclear industry is convinced that today's nuclear reactors are safe, and no radical design changes are necessary in the short-term future.[30] Nevertheless, the industry concedes that a number of design improvements will be necessary to make new nuclear reactors safer and more economical. The former consideration is not in conflict with the recognition that the industry believes that present reactors are safe. They do, however, recognize that in order to gain public acceptance, new reactors will have to be, or at least appear to be, considerably safer than present designs.

Advanced light water reactors (ALWR) will incorporate a number of these evolutionary improvements. For example, an ALWR was designed as a joint effort between Westinghouse, Mitsubishi and a number of other actors.[31] Its increased core size results in lower power density, and consequently decay heat would raise the core temperature more slowly than in a LWR. Emergency cooling is also made easier, extending the time operators would have to respond to emergencies.

The Electric Power Research Institute (EPRI) in the USA has developed a set of guidelines for the ALWR of the 1990s. Amongst other points, the ALWR should have a core-damage frequency of less than 10©5 events per reactor year, and a severe accident event of greater than or equal to 10©6 per reactor year, with a maximum dose rate of 25 roentgen within half a mile of the accident. In addition, the ALWR should be available at least 87% of the time, be refuelled every other year, and have a lifetime of 60 years. Furthermore, such a reactor should be constructed in 54 months. Finally, the ALWR should produce less than 70 m^3 of radioactive wastes, and cause less than 100 manrem of occupational exposure each year.[32]

Present day LWRs are far from these criteria. Estimates have been made for core-meltdown probabilities for precursor events in operating reactors, had they not been aborted by human or mechanical intervention. During the period 1969-79, this was 10©3 per reactor year, falling to 10©4 per year during 1980-81. Today's nuclear reactor's in the USA have been estimated to be of the order of 10©5, while some of the new designs are approaching the 10©6 per reactor-year value.[33] In the case of the Federal Republic of Germany a 1979 risk analysis calculated a probability of 10©4 of core meltdown per reactor year, while a more recent study in 1989 calculated 3.10©5.[34]

Average construction times of 54 months will also be a challenge. While Japan's average construction time has been consistently just above 50 months, other countries range typically between 70 and 100, with Brazil averaging 130 and India 161 months per reactor.[35] In the USA, construction times have risen from an average of less than 50 months to over 120 in the 1980s.[36]

Although the EPRI criteria are considerably more stringent than present averages, EPRI is convinced that such plants can in fact be built in the 1990s, based on the existing LWR technology, with appropriate improvement both in the technology and in the regulatory system. EPRI expects that the first ALWR will receive certification in the USA in 1990.[37]

The USA Department of Energy together with EPRI and a number of utilities from various countries are involved in a $ 200 million program to develop an ALWR (PALWR), incorporating passive core cooling features.[38] EPRI experts expect that such a reactor will receive certification in the USA by the mid 1990s. Medium- to large-scale HTGRs (High Temperature Gas-Cooled Reactor) are also being promoted by a number of countries, including the Federal Republic of Germany , USA, USSR, UK, Japan, and China.[39]

This evolutionary mode is the preferred way into the future of the nuclear industry in all the major nuclear-supplying countries. They have committed a considerable amount of human and financial resources to the LWR technology, and they do not see the need for any radical departures.

3.2 Liquid Metal Fast-Breeder Reactors

Fast-breeder reactors produce more fuel than they consume, by transforming natural uranium or thorium in their blankets to fissile elements. Experimental breeder reactors were built by a number of countries in the 1950s, expecting that uranium shortages in the future would require breeder technologies. These developments have not taken place. Uranium prices have not risen as expected, but the price of breeders has. Only a few countries continue active breeder programs.

The only commercial-size fast-breeder reactors are those in France and the USSR. In addition, the Federal Republic of Germany has also been building a prototype plant for 15 years, and Japan and India are still interested in developing their own breeder technology. Most of the experimental and prototype fast reactors have been beset by various difficulties, including problems with the liquid-sodium coolant.

The main reasons for the slowdown, or abandonment, of fast-breeder programs have been economics, safety, and in some cases, the issue of nuclear proliferation. The economics is particularly unfavorable to breeders. In France, for example, the Super Phénix is producing electricity at 2.5 times the electricity coming from traditional LWRs. In the international context, uranium prices would have to increase 3-6 times before breeders begin to show economies compared to the traditional LWRs.[40]

The future of breeder reactors is uncertain. It is likely that a number of countries presently involved in this technology will continue, but keeping to research, development, and prototypes. Even in the USSR and France, it does not seem likely that additional commercial-size breeders will be built during the coming decade.[41] While in the longer term a return of breeder technology is still possible (especially through newer, inherently safer designs, as discussed below), given the lead times necessary for development and commercialization, it is very unlikely to happen before well into the next century.

3.3 *Inherently Safe Reactors*

The term, "inherently safe reactor" (ISR) has been much discussed,[42] especially since the two major nuclear accidents at Chernobyl and Three Mile Island. Unfortunately, "inherently safe" seems to have different meanings for different people. A number of concepts need to be clarified.

In the traditional LWR, safety is achieved by add-on, engineered safety features, which include containment vessels of various kinds and sizes, cooling pumps, backups, control rods, and so on. The first reactors were made for military purposes, and neither safety nor cost was stressed as a design consideration. As they were scaled up into commercial units, these factors became increasingly significant. Particularly, as safety concerns intensified, progressively more safety devices had to be added on to the basic design, increasing both the costs and the complexity of the systems, the complexity itself leading to increased risk of accidents. Indeed, the present day LWR has been called "...inherently an exceedingly complex, unforgiving device."[43]

Absolute safety is not possible. There will always be some possibility that a particular event could occur. "Inherently safe" means that inherent to the particular reactor technology, basic physical properties of the system will ensure safe operation. Inherent safety, therefore, concerns the way safety is achieved, not the actual level of safety.

In ISRs the safety is achieved, not by engineered safety devices, as in the traditional LWR, but rather by relying on basic physical principles, such as thermodynamics and gravitation. For example, control rods can be inserted electromechanically, or cooling can be forced either by fans or pumps. In each case the electromechanical devices may fail due to power failures, earthquakes, chemical explosions, and so on, with potentially serious consequences. In a reactor with inherently safe design features, the control rods would fall into the reactor by gravity, and cooling would be via natural convection. Such features make reactors less complicated, smaller, and safer.

The question is how safe, given that absolute safety is impossible. How safe is safe enough for widespread public acceptance? Many reactor designers and analysts have addressed this issue. The crucial issue is whether it is possible to design and build a reactor, which under no conceivable scenario could release, in an uncontrolled way, important quantities of radioactive materials into the biosphere. The conceivable scenarios would need to include not only technical failure of the equipment, operator error, or earthquakes, but also terrorist actions whose aim might be to damage the reactor.

One environmental group in the USA has developed a set of criteria which an "acceptable" nuclear reactor should have. Such a reactor should have a core which will neither melt down nor catch on fire upon any conceivable operator action or sabotage, including abandonment of the site. Furthermore, such a reactor should be able to hold its radioactive inventory following attack with explosives (e.g., by a terrorist group) or an earthquake.[44]

If a reactor existed which could hold its inventory of radioactive materials within it, under *any* conceivable accident scenario, it could have a profound impact on the public perception of nuclear power. This may not be a sufficient condition to change the views of the critics, but it would be a necessary one. Furthermore, the safety of the reactor could be illustrated by creating "worst case scenarios," and *demonstrating* that the reactor can shut down and dissipate the remaining heat without any outside control.

Safety concerns have focussed on two aspects: quenching the nuclear reaction in the core, and dissipating the decay heat to avoid melting the core, which in most cases could result in major releases of radioactivity. Even if the reaction can be stopped easily, it has not been possible to design a conventional reactor where the heat could be dissipated under any conceivable circumstance.

Fuel integrity in the case of loss of coolant can only be achieved if the maximum temperature in the center of the reactor core is lower than the failure point of the fuel. The maximum temperature, however, is determined by the size of the reactor core, and the power density during normal operation. If the reactor is small enough, and the power density is low enough, then an inherently safe reactor design is achieved. The inherent limit seems to come to about 250-300 MWt, which is an order of magnitude smaller than present-day reactors.[45]

A number of reactor designs and even some prototypes exist which have inherently safe design features, some of which would probably pass the test of demonstrable safety. However, no such commercial reactors have yet been built, and even if they turn out in reality as good as on paper, they are a long way off as a major energy source.

One of the first inherently safe reactor designs was the PIUS (Process Inherent Ultimately Safe) reactor designed by the Swedish ASEA-ATOM.[46] PIUS is a 500 MWe PWR, whose core and primary cooling system is immersed in a large pool of borated water, housed in a large prestressed concrete vessel (PCRV). During normal operation the coolant pumps keep the borated water away from the core. Were these to fail for any reason, natural convection would carry the water to the core. The boron in the water would shut off the nuclear reaction, while the large volume of water would be able to cool the core for a number of days, without outside intervention.[47]

While the PIUS design has a number of inherent safety features, it is still dependent on some engineered features, such as the integrity of the large concrete vessel, the PCRV. Since it was first announced, improved designs based on the original PIUS have been presented in Sweden, Japan, and the USA,[48] but to date, the reactor has not yet been built. A demonstration plant could be built and operated in about a decade.[49]

The most promising reactor design, in terms of inherent and demonstrable safety is the MGR (modular high temperature gas-cooled reactor). Both Kraftwerk Union (KWU) of the Federal Republic of Germany, and General Atomic (GA) of the USA have proposed designs. The fuel is tiny spheres of uranium oxide encapsulated in successive shells of low density graphite, pyrolytic carbon, and silicon carbide, embedded in prismatic or spherical graphite fuel elements. The MGR's small power density and size inherently ensure that temperatures will always remain below 1600 °C, under any circumstances.[50] Intensive testing of the fuel has shown that they can perfectly contain the fuel up to about 1800 °C, and significant fuel failure occurs only at temperatures in excess of 2000 °C.[51] Since the MGR has an output of only typically 100 MWe, in practical applications a number of 100 MWe modules would probably be used together.

Although the basic principles of the MGR have been demonstrated in the Federal Republic of Germany, where a 15 MWe demonstration reactor was built in 1966 and operated until 1988, the reactor experienced a number of problems throughout its life.[52] A number of other designs have been developed by KWU, GA, as well as JAERI of Japan, using the basic principles outlined above. Nevertheless, no commercial reactor has yet been built.

An interesting combination of a modular gas-cooled reactor with a gas turbine (MGR-GT) has been proposed with a rated output of 200 MWt.[53] The system would be constructed from existing materials, within existing design codes, yielding 45-50% net efficiency at a cost substantially below that of steam generating plants. Such a plant is possible due to three important developments: (1) small modular reactors; (2) techniques of advanced heat exchangers and high efficiency turbomachinery now available; and (3) the advent of reliable, high efficiency solid-state power electronics.

One of the important advantages of the combination proposed above is its favorable economics. There are several reasons for such potentially favorable economics. The high thermal efficiency using the gas turbine reduces the costs per unit output. The small reactor size allows considerable factory production and assembly, making licensing simpler and faster, and reducing overall construction times. Finally, given the inherently safe features, a number of the complex, expensive, engineered safety devices will not be needed. Promoters claim that such a reactor could produce electricity for 4.5 c/kWh compared to clean coal at 5.5 c/kWh. Capital cost of "typical" US nuclear plants have been reported to be $2,300/kWe in 1987 dollars,[54] while the promoters of the MGR-GT claim cost to be $1000/kWe.[55]

The reluctance of the nuclear industry to invest in the development and demonstration of truly inherently safe nuclear reactors is perhaps the major reason why the technology is not taking off. One US commentator recently wrote "The American nuclear industry is its own worst enemy."[56] At the same time, the MGR is looked at as the possible savior of the nuclear option in the USA for the late 1990s and thereafter,[57] and the same view also can be found in Germany.[58]

Can the MGR make the difference for the future? Will the MGR be the "...pollutant free energy source"[59] of the future? While this reactor design certainly has advantages, such as improved safety, economics, and the possibility of a large export market, particularly to developing countries, the advantages have yet to be demonstrated. Claims for its demonstrable safety are questioned by some,[60] and will need to be tested. Furthermore, the actual systems costs can only be ascertained once a prototype commercial-scale reactor is built. Since the MGR is particularly suitable for the production of tritium,[61] it also will raise additional problems of nuclear proliferation.[62]

The IFR (Integral Fast Reactor) under development at the Argonne National Laboratory in the USA claims not only inherent safety, but also considerable reduction in the problems of nuclear waste, and improvement of the economics by carrying out fuel fabrication and a new, gas-phase separation reprocessing on the same site.[63] The IRF uses liquid sodium as coolant and a metal-alloy fuel. Since the latter is a good conductor, any excess heat is conducted to the large volume of coolant - the liquid sodium. Another advantage of the IFR would be that it would consume some of the actinides with very long half lives, which would be returned to the reactor after reprocessing. Argonne has tested an experimental, IFR-like reactor for loss of coolant in 1986, with positive results.

Nevertheless, the IFR has safety disadvantages. The main concern is the large quantity of liquid sodium which, if it comes into contact with water, is highly explosive, sufficient to destroy the reactor. Liquid sodium cooling is being experimented with in large-scale and experimental LMFBRs, as well as for solar towers. In both cases considerable problems have been reported,[64] including an important sodium explosion in the Almeria solar power station in Spain. As a result of this problem, a number of non-explosive coolants have been proposed, such as molten salts.[65]

3.4 Other Reactor Designs

There are a number of other new reactor designs which have inherent safety features. For example, Switzerland is particularly active in the development of small heating reactors of 10-50 MWt size, using principles of inherent safety, including low-core power densities and passive cooling systems. Other countries are also experimenting with such concepts, but no such reactors have yet been built.[66]

There has been a considerable discussion of completely different ways of producing nuclear reactors, in order to overcome some of the constraints to larger penetration of the nuclear option worldwide. Two such constraints are the large size of conventional reactors compared to the national grids of many countries, especially developing countries, and the various complications which arise during the very site-specific construction of each reactor. A number of proposals exist to scale down reactors to overcome the first constraint. Such scaling down would also allow the manufacturing and partial assembly of most components of the reactor at some factory location in one of the nuclear-supplying countries. The reactor could then be barge- or train-mounted and assembled on site, following basic civil-engineering preparatory work.

There would be a number of advantages, as well as disadvantages, of such a system. Factory construction and assembly could provide increased quality control, which would increase safety and overall availability factors, and thus improve the economics. On the other hand, dependence on foreign suppliers would increase. Given the strategic position of energy supply in all countries, such deals would only find limited applications.[67]

For the MGRs and the IFR prototype, demonstration will be necessary before either regulatory action can take place, and before the utilities are convinced that this is a technology which can be built and operated. This will require hundreds of millions of dollars of investment funds, probably from government sources. Actual ordering of such reactors could only take place subsequent to this, and is not expected until well into the first decade of the next century.[68] Consequently, without major intervention from nuclear supplier countries, such as the USA, FRG, etc., and intensive international cooperation, it is unlikely that inherently safe nuclear reactors will make an important penetration of the energy-supply sector during the period under consideration in this study. However, taking a longer perspective, it is probable that during the first two decades of the next century, increasingly reactors with inherent safety features will be the reactors of choice.

As for the level of safety provided by these new reactor designs, again, only full demonstration will tell, and only in such a case can these reactors make any impact on public opinion. While the demonstration of such safety would give a major boost to the nuclear industry, other public concerns would still need to be addressed.

4. The Disposal of Radioactive Wastes

Public concern over radioactive waste is a major impediment in many countries. In a number of industrialized countries, legislation has been introduced which ties the continued operation and further construction of nuclear-power plants to the resolution of the issue of nuclear-waste disposal.

"Nuclear waste" refers either to spent nuclear fuel which is to be disposed directly, or reprocessed waste of various categories.[69] Depending on the country, reprocessed wastes are categorized differently, but generally some kind of low-, medium- (or intermediate-) and high-level wastes are differentiated. We focus on high-level waste here, since it has the most difficulties. It is assumed that if the capability exists to safely and convincingly dispose of high-level wastes, all other radioactive wastes can also be handled.[70]

A related problem is the decommissioning of nuclear reactors. The decommissioning process would produce large quantities of low- and medium-level wastes, but the quantity of high-level wastes will not be significantly affected. Since no large-scale commercial reactor has yet been decommissioned, the costs, the technical details, the problems, and the quantities and types of wastes which will result, are not known with certitude. Decommissioning costs have been estimated at around 200 $/kW, although this is probably an underestimate.[71] Assuming this optimistic figure, the total costs of decommissioning are not expected to make nuclear power considerably more expensive, given other fuel-cycle costs.[72] Were decommissioning costs to rise significantly higher than presently estimated, the high cost could act as additional deterrent to new plant ordering. Furthermore, almost irrespective of price, given the high volume of wastes, the decommissioning process is going to concern public opinion, similar to the overall problem of waste disposal.

4.1 *The Size of the Problem*

Considerable quantities of nuclear wastes have already been produced, and they will have to be disposed of, one way or another. Although the total amount of nuclear waste generated by the nuclear fuel cycle is small, compared to the quantities of hazardous wastes produced by industry, its high radiotoxicity requires its isolation from the biosphere. The total amount of spent fuel generated by nuclear energy worldwide, up to 1988, was 15,100 m^3, which is expected to rise to 24,700 m^3 by 2005. In addition, some of the spent fuel is reprocessed, resulting in high-, medium- and low- level wastes. The high-level wastes totalled 3,200 m^3 in 1988, and are expected to rise to 5,120 m^3 by 2005.[73]

During the last decade, considerable progress has been made in the quest for the disposal of radioactive wastes. Low- and medium-level wastes are being disposed in a number of countries, and plans are under way for high-level wastes also.[74] However, to date no high-level waste (or spent nuclear fuel) has been disposed of, in a final repository, and all is now stored in various forms of temporary facilities, either on reactor sites, or in specially designed storage facilities.

4.2 *Problems of Geologic Disposal*

It is difficult to predict what will happen during the lifetime of a repository. Anticipating all possible human and natural actions, which could compromise the safety of a repository over a few hundred thousand years, is daunting indeed. While some of the biological and geophysical processes which have impacts on a repository can be modelled fairly well, it is extremely difficult to envisage the full spectrum of possible scenarios, and it is practically impossible to assign probabilities to inadvertent or intentional human intrusion into a repository.[75]

For example, one of the criticisms of the technically advanced Swedish waste-disposal concept underlines this point. In this concept, the spent nuclear fuel is encapsulated in thick pure copper canisters, which are then placed in a deep granite formation. The estimated life of the canisters is in the millions of years. Yet, a few thousands of years from now, in a completely different generation, when all the records of this waste repository are perhaps lost, people may find a fantastic, pure copper mine there, and decide to mine the resource.

There is no "proof" of a successful repository in the traditional sense. The time horizons are so long that national boundaries, civilizations, and indeed everything, will change in that period of time. This is recognized by promoters and critics alike. The promoters are searching for various ways of demonstrating safety, by combining various modelling techniques, and some hands-on experimental methods. The critics seem to be unanimous that the institutional process used in safety assurance, and not the technical demonstration, would be the critical factor. Overcoming the scientific uncertainties of the disposal process will require institutional processes which inspire confidence in those who certify that the safety has been met.[76]

One technical solution to the extremely long time horizons involved would be to "burn up" the long-lived actinides in nuclear reactors, in a process called transmutation. Following neutron irradiation in a nuclear reactor, the long-lived nuclides would transmute into shorter-lived ones with half lives of the order of hundreds of years and less, instead of the tens of thousands of years for some of the actinides. This would correspondingly reduce the geological waste-disposal problem. Although the physical feasibility of nuclear transmutation is not in doubt, no engineering feasibility study has been carried out,[77] and presently no country is actively pursuing this option. Major disadvantages of the transmutation concept include the substantial costs which may be involved, and the fact that it requires considerably more refined reprocessing of the spent fuel.

The siting of a nuclear repository is a delicate socio-political process. The not-in-my-backyard (NIMBY) syndrome is evident.[78] In addition to technical criteria for site selection (which tend to be fairly uniform in countries which have waste-disposal programs), countries are increasingly using non-technical criteria, which are better categorized as criteria of "political acceptability."[79] Thus many of the potential or planned sites for repositories have been selected near existing reactors or other nuclear-facility sites. Similarly, many sites were chosen in depressed local communities, where the economic compensations (e.g., taxes or outright compensation payments) would pay off in terms of public acceptance.

In Sweden the relatively successful waste-disposal program is due, at least in part, to the 1980 decision to phase out nuclear energy in the 1990s. Consequently, the opposition to nuclear power does not see the waste-disposal program as a *de facto* acceptance of the nuclear program.[80]

4.3 *Transboundary Movement of Radioactive Wastes*

There has been considerable discussion in the literature about the desirability and the possibility of an international repository for high-level radioactive wastes.[81] The rationale behind such a concept is partly technical and partly political. On the one hand, some of the most stable geological formations of the world are not where population densities are high (and where nuclear power is intensively used), but often in remote areas, which are either international territory (e.g., sub-seabed, Antarctica), or are in remote parts of national territories (e.g., N. Canada, N. E. Siberia). Technically, the location of large repositories in such areas would be ideal. Such solutions are particularly favored by small nations, whose nuclear programs are too small for independent work on a repository to be economical. Furthermore, many of these countries either may not have appropriate geologies for waste disposal, or may simply be concerned because of high population densities.

The Nuclear Energy Agency of the OECD has been coordinating an international program on sub-seabed disposal. To date, technical feasibility of such a concept has not yet been provided. Politically, such a solution would be even more difficult, since a number of international conventions would prohibit such an act. Nevertheless, work continues on the technical feasibility of the sub-seabed option. Most observers, however, exclude the possibility of an international repository before national repository programs begin to take more concrete shape in the next century.

Countries which are interested in disposing of their nuclear wastes outside their own territories have apparently initiated contacts with developing nations to take these wastes for a fee. While a number of reports have appeared in the international press describing such negotiations, no agreements have yet been made, and officials of both exporting and importing countries tend to deny them.[82] Although presently not contrary to international law, such actions could reinforce a dangerous trend toward the export of pollution to developing countries. The report of the World Commission on Environment and Development also called attention to this problem, and called for international agreements to regulate such activities.[83]

During 1988 the dumping of radioactive and toxic waste was reported in the international press.[84] In the reported cases, either the importing country was unaware of the radioactive content of these wastes, or they "neglected" them. Investigations of a number of these claims found considerable quantities of toxic, but no radioactive, wastes.[85] In the meantime, the IAEA has set up a committee to advise its Secretary General on this matter. It is expected that this committee will draft guidelines on the transboundary movement of all radioactive wastes, which eventually could be drafted into a binding international agreement.

4.4 *Discussion*

With considerable difficulty, national programs for final radioactive waste disposal are progressing. It is unlikely that an international repository will be available until well into the next century when the first national final repositories will have been filled and closed. Consequently, all final disposal will initially take place in national repositories. Although the problems of site selection and public acceptability will probably intensify as nations get closer to actual disposal, based on progress to date, it is not likely that the final disposal of high-level radioactive waste will be the major limiting constraint on the further development of nuclear power. The required repositories will probably be built out of sheer necessity. In a number of countries, however, public opposition to geological waste disposal may substantially slow nuclear-power programs.

5. Nuclear Proliferation

This issue, although less discussed now in the popular and scientific literature than in the 1970s, has not disappeared. With increasing numbers of nuclear plants, more nuclear fuel, spent fuel, and reprocessed wastes will be transported both within and between countries; hence proliferation concerns will increase. While some of the nuclear technologies are technically less prone to proliferation issues, no nuclear fuel cycle under discussion is without problems.[86]

A typical 1 GWe LWR produces approximately 250 kg of plutonium annually. Following reprocessing of the spent fuel, the recovered plutonium is sufficient to make 25 nuclear warheads. The plutonium can also be mixed with natural uranium to form mixed oxide (MOX) fuel. Although the latter is not weapons-grade fissile material, a low-yield nuclear explosive in the kiloton range can be constructed from it. Canada, the USA, and Sweden have made formal decisions not to reprocess spent fuel. All other nuclear countries are reprocessing (or planning to reprocess) at least part of their spent fuel. Countries like France, the UK, Germany, and Japan have made considerable technological and financial investment in their reprocessing activities.[87]

As a result, there will be considerable quantities of plutonium and mixed oxide fuel transported between and within countries by the turn of the century. The four major reprocessing countries will be separating about 27 tonnes of plutonium per year by the end of this century, resulting in hundreds of shipments containing plutonium per year.[88] Terrorist groups could divert some of this material and prepare crude, but functioning, nuclear explosives. Also, individual nations could use their nuclear fuel-cycle programs to help them reach a nuclear weapons capability, through acquiring general nuclear experience, or switching from civilian to military activities once the program is under way.

Thus individual countries or terrorist groups may divert weapons-grade or weapons-usable nuclear materials. Avoiding this risk will require considerably increased controls, causing possible loss of liberties, proliferation of risks, health and environmental damage, and considerable increase of public opposition to nuclear energy worldwide.

6. The Public Acceptance of Nuclear Energy

The public voice has become increasingly strong and negative in most industrialized countries on nuclear energy. In the USA during the ten-year period starting in 1975, the fraction of respondents who opposed nuclear power increased from about 15% to 75%, while the fraction who favored it decreased from about 65% to 15%.[89] Similar tendencies can be observed in other industrialized countries.[90] In Japan, although a long history of public opposition exists, it has not been sufficiently strong to be a major obstacle to the nuclear program.[91]

More recently, public attitudes toward nuclear power have been influencing nuclear-energy policies of centrally planned and developing countries as well. Public opposition has contributed to scaling down or stopping nuclear developments in countries like Brazil, Mexico, and the Philippines.[92] In the USSR, opposition to nuclear power has been publicly voiced following the policy of *glasnost*. There seems to be some form of public opposition to virtually all new plants now under construction.[93]

The public's fear of, and attitude to, nuclear energy is a complex issue.[94] Important factors which seem to affect public opinion include the connection to military technology; the perceived costs and benefits; the issue of safety; as well as questions of philosophy and life styles. The first large-scale use of nuclear science was for military purposes. Present-day nuclear reactors were developed directly from US military submarine technology. Also, the science and technology required for civilian nuclear power can be used for military purposes. There is considerable evidence that nuclear power has contributed to military applications and may do so, in the future, in those countries which do not yet possess such weapons.

The connection with the military has deeply affected public perception. Many people have never been able to dissociate civilian reactors from the Hiroshima and the Nagasaki bombs. Often, nuclear energy *per se*, is thought to be evil. More practically, the fear of civilian technology giving rise to military applications, has shaped important decisions relating to the use of nuclear energy. Examples are the US decision not to reprocess spent nuclear fuel in the 1970s, and the restriction of the supply of nuclear technologies and fuels to certain countries, particularly those which have not signed the Non Proliferation Treaty. Reports of the Nukem/Transnuclear scandal from the Federal Republic of Germany,[95] as well as disappearance of civilian plutonium and its possible use in military applications in the UK,[96] continue to reinforce public perceptions of the connections between military and civilian nuclear technologies.

A crucial factor shaping public attitudes has been the costs of nuclear power, compared to the benefits it provides. Nuclear costs have been very difficult to measure, since the calculation needs to include a number of hidden costs, such as subsidies. In countries such as France or Japan, where the relative cost of nuclear power is perceived to be low, and its benefits great, the opposition tends to be less. The converse is true in the USA, where fossil-fuel-based power is relatively cheap and easily available, while the costs of nuclear are high.

The costs of nuclear energy, however, include more than just the narrow economic costs. Nuclear risk is another "cost" which has had an important effect on public opinion. These risks can be grouped into two categories: major nuclear accidents resulting in important releases of radioactivity, or chronic, low-level radiation releases. Major releases can occur from different parts of the nuclear-fuel cycle, such as reactors, reprocessing plants, storage and transportation facilities. While the probability of large accidents is claimed to be very small, the consequences can be very high. In such cases the public tends to de-emphasize the low probability and emphasize the possible consequences.[97]

The most serious nuclear accident so far was in Chernobyl in the USSR in 1986. It has had a profound impact on the public's view of nuclear power worldwide. In addition to the major release of radioactivity, perhaps the most important dimension of this accident was its international character. The radioactive release affected all of Europe. Although in all countries the sharp rise in the percentage of persons opposed to nuclear power has decreased somewhat since, in none has it reached its pre-Chernobyl level. More detailed analysis shows that public response in these countries was directly proportional to the average 50-year whole- body dose received in those countries. Furthermore, in those nations with a high percentage of citizens uncommitted in their view for or against nuclear power, the initial increase in opposition was large, and relatively permanent.[98]

The nuclear industry has learned a great deal from these accidents, both in the sense of improving technical safety, and in terms of organizing emergency response.[99] Through the efforts of various international organizations, and particularly the IAEA, the international community is much better prepared for another major nuclear accident with transboundary effects.[100] Nevertheless, another major nuclear accident would certainly further increase opposition of the public, and could possibly provide the final blow to increased nuclear-energy utilization in a number of countries.

Another major safety issue which causes public concern is the controversy surrounding the effects on humans of low-level radiation. Low-level releases can occur from the routine operation of fuel-cycle facilities or from the fallout following major nuclear accidents or a breached waste repository. There is no clear scientific consensus on the effects of low-level radiation on the human body.[101] For example, the long-term additional cancer deaths due to the Chernobyl fallout have been estimated to lie anywhere between 0 and 28,000 additional deaths over the next 50 years, depending on which assumption one believes for the health effects of low-level radiation.[102]

Finally, many people are opposed to, or are concerned about, the use of nuclear energy for reasons of life-style, philosophy, or even religion.[103] Nuclear energy may be considered inherently evil or bad, while for many people it may be the ultimate symbol of the materialist, ever-growing western industrialized mode of society. In such cases opposition to nuclear energy can be one means by which other social goals can be achieved.

Often proponents of nuclear energy try to brush aside the above issues, labelling them "irrational, unfounded, and unscientific." It must be recognized, however, that the public perception of technological risk (and not just of nuclear power, although this certainly represents technological risk *par exellence*) does not obey simple, scientifically determined equations where one inputs the probability of a reactor meltdown and then obtains a clear indication of public attitude. Public attitudes are the results of a complex interaction of information provided by the scientific community, cultural, religious and psychological issues, and other socio-political values of the public.[104]

There are no easy, generic solutions for improving the public's view of nuclear energy. However, a few issues are clear from the various studies which have been carried out on this subject in the past. First of all, nuclear energy has to be perceived as having *much* lower cost-benefit ratios than the available alternatives. This means that the overall system-price for nuclear energy must go down considerably. Slightly lower than conventional alternatives will not suffice. This would be one advantage of the MGR-GT concept discussed in Section 3, which claims to do just that.

But the benefits side must be equally clear. In order for any major increase in public acceptance of nuclear energy to take place, the public must be convinced that nuclear energy is clearly the *only* alternative to produce the required energy. So far, the public is not convinced of this. Will the climate-change issue be sufficient to change the view of the public? It is difficult to predict this. Nuclear-power advocates will have to try to present the case that nuclear energy is the quickest, cheapest, and socially and politically most acceptable way to reduce greenhouse-gas emissions. So far this has not yet happened and, if anything, the contrary seems to be true.[105]

7. The Future of Nuclear Power

A study has been prepared by the Nuclear Energy Agency (NEA) of the OECD about the short- and medium-term prospects of the role of nuclear energy in reducing greenhouse-gas emissions in the OECD countries. According to this study,[106] the use of nuclear energy has already helped to "avoid" considerable quantities of carbon- dioxide emissions in the electricity sector, and this is expected to grow at a modest pace during the next two decades. However, overall, carbon-dioxide emissions will continue to grow.

According to the OECD/NEA projections, nuclear electrical generating capacity will climb to 320 GWe by 2005, representing about 17% of total electrical capacity during the next two decades. The existing nuclear generation in 1987 "avoided" 0.3 Gt of carbon-dioxide emissions, had those been constructed as coal-based power plants. This should be compared to total OECD carbon-dioxide emissions of 2.4 Gt in 1985.[107] By 2005 the total OECD carbon-dioxide emissions will have risen to about 4 Gt,[108] and nuclear energy will have avoided an additional 0.2 Gt. While this is not negligible, its total impact on greenhouse gases would be marginal at best.

It is also important, however, to look at a longer-term perspective, as was done in another recent OECD/NEA study.[109] The study assumes that socio-political and economic constraints to the further growth of nuclear energy are removed,[110] and there is a moderate environmental push for increased nuclear energy.[111] The study finds that under such assumptions, it might be conceivable that OECD nuclear generating capacity would increase up to a maximum of 500 GWe by 2005 and 1120 GWe by 2030. Even such an optimistic scenario would result in only about half a gigatonne of additional reductions by 2005.[112]

This very optimistic scenario for nuclear development, however, would not be easy to achieve. OECD countries, on the average, would have to reach 70% of their electricity from nuclear on an asymptotic curve, with a halving time of only 20 years, involving the construction of 27 GWe nuclear capacity every year. While this is not infeasible from a technical point of view, a number of constraints appear.

Installed nuclear capacity in the OECD countries grew at rate of around 17 GWe between 1975-1985. The expected rate during the present period of 1985-2000, however, is only 6.5 GWe, coming down to around 3 GWe/yr during the last decade, 1990-2000. If orders started coming in increasing numbers, the industry could probably gear up to the 27 GWe/yr target sometime in the next century. The likelihood of this happening, however, is very small, given the constraints described above. Even a 10 GWe/yr rate may be difficult to achieve. Additional problems relating to the decommissioning of nuclear plants, of manpower availability, sufficient fuel-cycle capacities, and even the availability of uranium resources might present constraints to such developments.

The official COMECON forecasts for the centrally planned economies in Europe are high, and expected to increase capacity from 30 GWe presently to 224 GWe by 2000,[113] but it is questionable if these can be realized, given the present climate of increasing economic difficulties and increasing opportunities for public opposition to develop. The IAEA projects a more modest growth of 120 GWe by 2005.[114] Using the IAEA projections, and the same emission factors as in the OECD study, we find that by 2005, nuclear power in the CPE countries could contribute to the avoidance of only about 0.3 additional gigatonnes of carbon per year. If we assume the more optimistic COMECON figures, the avoided emissions per year are still only about 0.6 Gt.

Apart from a few success stories, mainly limited to the newly industrializing countries (NICs), nuclear power has not managed to substantially penetrate into the electricity supply of developing countries. The IAEA estimates an increase to 50 GWe capacity by 2005, while Surrey (1988) estimates 30 GWe by 2000.[115] These projections are probably an upper limit of what may take place by the turn of the century; given the prospects for overall development, and a resolution of the debt crisis, the situation is unlikely to change radically early in the next century.

Assuming the total projected nuclear capacity for the developing countries to be around 30 GWe by the end of this century, and assuming emission and substitution factors as in the OECD study, the total avoided yearly carbon-dioxide emissions would come to about 0.1 gigatonne. Even assuming (arbitrarily) a doubling of this capacity, the emission reduction would still only be about 0.2 Gt per year.

8. Conclusions

While nuclear energy has its problems, it is here to stay. It is unlikely that it will die out completely. Even in a worst-case scenario for nuclear development, there will continue to be countries that will wish to keep a certain level of nuclear generation.

It is equally unlikely that, given the various constraints on nuclear-power growth outlined in this chapter, a major expansion of nuclear power will be possible during the next two decades. The availability of new technologies, such as inherently safe reactors are not likely to influence this much. The most likely situation for the next two decades, therefore, is a continuation of events: modest overall nuclear growth, with considerable inter-country and inter-regional variations, more or less following the IAEA projections cited in this study, as the upper bound. In such a case, the nuclear option could be responsible for approximately 5-25% of yearly carbon-dioxide emission reductions by 2005, with the higher figure representing the effect of nuclear-energy use in a "high energy efficiency" scenario, where a much greater quantity of carbon-dioxide emissions are avoided by the efficiency measures themselves.

Given even the most optimistic future for nuclear power expansion, the upper limit of the avoided yearly carbon emissions three decades from now will be somewhere around 1.7 gigatonnes, representing between ten to thirty percent of total emissions. While this is not negligible, it is nowhere near the required targets, and it should be looked at as an upper bound. The likelihood of such an "optimistic" future is very low given the constraints outlined in this chapter.

If the opposition to nuclear power turns out to be more persistent, even the "modest" IAEA projections will not be realized, and even if they did, the rates of nuclear penetration could not be sustained much beyond that, since replacement of old nuclear plants in such a climate would not be possible. In such a "pessimistic" scenario, the contribution of nuclear power to carbon-dioxide reduction strategies could go down to zero by the end of the second decade of the next century, assuming that none of the presently functioning reactors are replaced by nuclear units. This "pessimistic" scenario, however, seems equally unlikely as the very "optimistic" scenarios discussed above.

The most important conclusion, therefore, is that although nuclear energy is a low carbon-dioxide energy system, it is by no means a very good tool for reducing the targeted carbon emissions by the required time. This is because carbon-dioxide emissions are not only from the electricity sector but also from transport, industry, and the domestic sectors, where nuclear has less possibilities for substitution. Furthermore, given long planning and construction times, nuclear development takes a very long time. Closing down existing nuclear reactors for whatever socio-political goals, however, would imply that much more will need to be done to reduce carbon-dioxide emissions from other demand or supply-side strategies. From the point of view of climate change, replacing existing nuclear reactors with fossil fired ones would be stepping backwards.

This then leads us to the most important next step. If not nuclear energy, than what are the best tools for reducing carbon-dioxide emissions during the next three decades? While such a comparative study was not the subject of this paper, there is sufficient evidence accumulating that energy-efficiency measures are indeed more effective. They are cheaper, and they can be implemented much faster than increasing nuclear supply.[116]

While the scope for efficiency measures is very large in all countries, at some point all potential efficiency measures will have been exhausted, and new supplies will be necessary. In many developing countries, with or without strong efficiency measures, additional supplies may be necessary at earlier stages. At that point, nuclear energy will need to compete with the available options at that time, which will probably include new supplies from hydro, solar, other renewables, and some advanced fossil systems.

Notwithstanding the various problems of safety, waste disposal, nuclear proliferation, and the overall economics, we should not foreclose the option that nuclear energy, particularly in the context of a "second nuclear era," could be one element of that future supply mix. Where additional supplies are needed, the nuclear risk can be perhaps considered worthwhile, given the risk of global climate change.

The global warming problem, as important and serious as it may turn out to be, does not justify a considerably increased global nuclear program for the next two to three decades. Even if, for other political or socio-economic reasons, such an intensive global nuclear program were initiated, its impact on carbon-dioxide emissions would be only marginal. This conclusion is true whatever may be the costs and feasibilities of alternative emission-reduction strategies, such as energy-efficiency measures, or the availability of other low carbon-dioxide energy supplies, such as additional hydro, direct and indirect solar, and some biomass options.

Table 2: Present and Projected Contribution of Nuclear Power to Carbon-Dioxide Reductions, Using the 1989 IAEA and the Optimistic Projections.

	1987			2005-IAEA			2005-Optimistic		
	a.	b.	c.	d.	b.	c.	d.	b.	c.
OECD	2.4	0.3	-	4	0.5	0.2	4	0.8	0.4
CPEs	1.3	0.1	-	4	0.4	0.3	4	0.7	0.6
DCs	1.4	0.05	-	2	0.1	0.1	2	0.2	0.2
World	5.1	0.45	-	10	1.0	0.6	10	1.7	1.2

NOTES: The numbers in this table should be interpreted with caution, and only as indicative, particularly since the carbon-dioxide projections shown are not necessarily consistent with the nuclear projections.

a. total carbon-dioxide emissions in gigatonnes of carbon for the year 1985 (WRI 1988)
b. yearly avoided carbon emissions in gigatonnes due to nuclear power use
c. incremental yearly avoided carbon emissions (after 1987) in gigatonnes
d. projected total yearly carbon emissions in gigatonnes (Edmonds & Reilly 1985)[117]

Notes

1 Environment Canada, *The Changing Atmosphere: Implications for Global Security*, Conference Statement, Toronto, Environment Canada, June 1988; hereafter cited as: Toronto 1988. The United Nations Environment Programme (UNEP) and the Beijer Institute (BEIJER), *The Full Range of Responses to Anticipated Climatic Change*, Nairobi, Kenya, UNEP, April 1989; hereafter cited as: UNEP and BEIJER 1989. World Meteorological Organization (WMO), *Report of the International Conference on the Assessment of the Role of Carbon Dioxide and of Other Greenhouse Gases in Climate Variations and Associated Impacts*, WMO-No.661, Geneva, Switzerland, WMO, 1986. World Meteorological Organization, *Developing Policies for Responding to Climatic Change*, WMO/TD-No. 225, Summary, Geneva, WMO, April 1988. Woods Hole Research Center, *International Conference on Global Warming and Climate Change: Perspectives from Developing Countries*, Conference Statement, Massachusetts, Woods Hole Research Center, February 1989.

2 The use or non-use of nuclear energy will have small direct and indirect affects on the emissions of other greenhouse gases which are neglected in this study.

[3] World Resources Institute and the International Institute for Environment andDevelopment, *World Resources 1988-89*, New York, Basic Books, 1988; hereafter cited as: WRI 1988.

[4] Jose Goldemberg, Thomas B. Johansson, Amulya K. Reddy, and Robert H. Williams, "An End-Use Oriented Global Energy Strategy," *Annual Review of Energy* 10 (1985): 613-688.

[5] Jae Edmonds and John M. Reilly, *Global Energy, Assessing the Future*, Oxford, Oxford University Press, 1985; hereafter cited as: Edmonds and Reilly 1985.

[6] Bill Keepin, "Review of Global Energy and Carbon Dioxide Projections," *Annual Review of Energy* 11 (1986): 357-392.

[7] This is somewhat arbitrary, but does represent reasonable bounds of the large scatter in the various projections. These bounds should not be interpreted as targets. Several studies have indicated that a halving of carbon-dioxide emission rates by 2020 would be necessary to avoid disastrous climate change.

[8] Toronto 1988.

[9] UNEP and BEIJER 1989.

[10] Timothy Jackson, "Is Fusion Feasible? An Assessment of the Methodology and Policy Implications," *Energy Policy* 17, No.4 (August 1989): 407-412.

[11] Meridian Corporation, *Energy System Emissions and Material Requirements*, Study prepared for the US Department of Energy, Draft Version, Washington, D.C., Meridian Corporation, November 1988.

[12] Renewable alternatives have considerably lower system-emissions than nuclear. For example, photovoltaic systems emit about one tenth and small-scale wind about one fifth of carbon-dioxide emissions from the nuclear cycle, according to figures in U. Fritsche, L. Rausch, and K. H. Simon, *Environmental Analysis of Energy Systems: The Total-Emission-Model for Integrated Systems*, Darmstadt, Federal Republic of Germany, The Oeko Institute, September 1989.

[13] International Atomic Energy Agency (IAEA), "Nuclear Power and Fuel Cycle: Status and Trends 1989," in IAEA, *IAEA Yearbook*, Part C, Vienna, IAEA, August 1989; hereafter cited as, IAEA 1989a.

[14] IAEA 1989a.

[15] IEAE 1989a.

[16] World Energy Conference, *Global Energy in Perspectives 2000-2020*, 14th Congress, Montreal, Canada, World Energy Conference, 1989.

[17] Nuclear Energy Agency (NEA), *Nuclear Energy Data-1989*, Paris, NEA/OECD, 1989; hereafter cited as: NEA 1989.

[18] NEA 1989.

[19] NEA 1989.

[20] Asa Moberg, *Nuclear Power in Crisis*, Amsterdam, WISE, 1986.

[21] NEA 1989.

[22] International Atomic Energy Agency, *Nuclear Energy and the Environment: A Report Prepared at the Request of the IAEA General Conference for Presentation*

to the United Nations General Assembly, Vienna, IAEA, May 1989; hereafter cited as: IAEA 1989c.

[23] John Surrey, "Nuclear Power: An Option in the Third World?" *Energy Policy* (October 1988); hereafter cited as: Surrey 1988.

[24] Lu Yingzhong, *Long-Term Energy Forecasting of PRC and Strategies in the Development of Nuclear Energy, P.O. Box: 1021, Beijing, China, Institute of Nuclear Energy Technology*, December 22, 1986.

[25] IAEA 1989c.

[26] Surrey 1988.

[27] Lawrence M. Lidsky, *Modular Gas-Cooled Reactors for Electric Power Generation*, Paper presented at the International Nuclear Engineering Symposium on the Development and Use of Small and Medium Size Power Reactors in the Next Generation, Japan, Tokai University, November 19-20, 1986; hereafter cited as: Lidsky 1986.

[28] Alvin M. Weinberg and Irving Spiewak, "Inherently Safe Nuclear Reactors and a Second Nuclear Era," *Science* 224 (June 29, 1984): 1398-1402; hereafter cited as: Weinberg and Spiewak 1984.

[29] For a detailed study of different advanced reactor technologies, see International Atomic Energy Agency, *The Nuclear Fuel Cycle System, A Directory of Nuclear Facilities*, IAEA-TECDOC-479, Vienna, IAEA, October 1988.

[30] Chauncey Starr, *Reshaping Fission for Social Acceptability*, Paper presented to the 11th International Scientific Forum on Fuelling the 21st Century, Moscow, September 29 - October 6, 1987. J. J. O'Conner, Chief Executive Officer of Commonwealth Edison, Testimony to the US Senate Nuclear Power Oversight Committee, May 1988. K. E. Stahlkopf, J. C. DeVine, and W. R. Sugnet, *US ALWR Programme Sets Out Utility Requirements of Future Plans*, 1988; hereafter cited as: Stahlkopf, DeVine, and Sugnet, 1988.

[31] Richard K. Lester, "Rethinking Nuclear Power," *Scientific American* 254, No. 3 (March 1986); hereafter cited as: Lester 1986.

[32] Stahlkopf, DeVine, and Sugnet, 1988.

[33] Alvin M. Weinberg, "A Nuclear Power Advocate Reflects on Chernobyl," *Bulletin of Atomic Scientists* 43, No. 1 (August/September 1986).

[34] Oeko Institute, *Umwelt-, Sicherheits-, Entsorgungs- und Akzeptanzaspekte der Kernenergienutzung*, Personal Communications from the Oeko Institute, Freiburg, Federal Republic of Germany, Oeko Institute, October 1989; hereafter cited as: Oeko 1989.

[35] International Atomic Energy Agency, *The Effects on the Environment of the Dumping of Nuclear....*, Presented to the Governing Council of the United Nations Environment Programme, Nairobi, Kenya, IAEA, May 1989; hereafter cited as: IAEA 1989b.

[36] Lester 1986.

[37] K. E. Stahlkopf, Electric Power Research Institute (EPRI), Personal Communication, Palo Alto, California, September 20, 1989; hereafter cited as: Stahlkopf 1989.

[38] See the section on Inherently Safe Reactors for the definition of passive features.

[39] Oeko 1989.

[40] Oeko 1989.

[41] Oeko 1989.

[42] Weinberg and Spiewak 1984. Irving Spiewak and Alvin M. Weinberg, "Inherently Safe Reactors," *Annual Review of Energy* 10 (1985): 431-462. Lu Yingzhong, *Ordeals of Chernobyl and the Re-justification of the Inherently Safe Reactors*, Paper presented to the 11th International Scientific Forum on Fuelling the 21st Century, Moscow, September 29 - October 6, 1987; hereafter cited as: Lu 1987. Jack N. Barkenbus, "Prospects for Inherently Safe Nuclear Reactors," *Energy Policy* (February 1988).

[43] Lawrence M. Lidsky, "Safe Nuclear Power," *The New Republic* (December 28, 1987); hereafter cited as: Lidsky 1987.

[44] Jan Beyea, National Audubon Society, *Is There Any Role for Nuclear Power in Preventing Climate Disruption?* Testimony before the US House Subcommittee on Energy and Power, USA, March 15, 1989; hereafter cited as: Beyea 1989.

[45] Lawrence M. Lidsky, *Nuclear Power, Different Stokes*, Paper presented at the September 7, 1988 meeting of the Environmental Dialogue Group of the Edison Electric Institute, USA, Edison Electric Institute, 1988.

[46] Now merged with Brown Boveri of Switzerland, forming the ABB (Asea Brown Boveri) multinational.

[47] Weinberg and Spiewak 1984.

[48] ASEA-ATOM has proposed PIUS-II, while IHHI of Japan designed the ISER-1, ISER-2, and ISER-HV, and ORNL in the USA proposed the 750 MW PIUS-BWR (Lu 1987). Kaere Hannerz, "Making Progress on PIUS Design and Verification," *Nuclear Engineering International* (November 1988).

[49] Weinberg and Spiewak 1984.

[50] Larger high temperature gas-cooled reactors (HTGR) have also been proposed in a number of countries, including the USA and the FRG. However, due to the increased core sizes, some of the inherently safe design features are lost in these concepts. Consequently, the reference concepts show a tendency toward the smaller designs (Oeko 1989).

[51] Lidsky 1987.

[52] Lothar Hahn and Britta Nockenberg, *Entwicklungsstand und Probleme von Hochtemperaturreaktoren-eine Argumentationshilfe*, Freiburg, Federal Republic of Germany, Oeko Institute e.V., Mai 1989; hereafter cited as: Hahn and Nockenberg 1989a.

[53] L. M. Lidsky, D. D. Lanning, J. E. Staudt, X. L. Yan, H. Kaburaki, and M. Mori, *A Direct-Cycle Gas Turbine Power Plant for Near-Term Application-MGR-GT*, Presented at the 10th International HTGR Conference, San Diego, California, September 19-20, 1988. M. Mori and L. M. Lidsky, *Recuperation Design for a Nuclear Brayton Cycle Power Plant*, Cambridge,

Massachusetts, Department of Nuclear Engineering, Massachusetts Institute of Technology, 1988.

54 The "typical" reactor-cost estimate needs to be compared to $5-6000/kWe for the Seabrook nuclear plant in Maine and the Shoreham reactor in Long Island, USA. Charles Komanoff, *Variations in Nuclear and Coal Plant Capital Costs*, New York, KEA, November 13, 1989.

55 Paul E. Gray, "Nuclear Reactors Everyone Will Love," *Wall Street Journal* (August 17, 1989); hereafter cited as: Gray 1989.

56 Gray 1989.

57 Lidsky 1986.

58 Lothar Hahn, *Der Kleine Hochtemperaturreaktor-Letzter Strohhalm der Atomindustrie?* Frieberg, Federal Republic of Germany, Oeko Institute e.V., Maerz 1988; hereafter cited as: Hahn 1988.

59 Colin F. McDonald, "Nuclear Power--A Technology Vector in Carbon Dioxide Emission Reduction and Acid Rain Abatement," in *Small and Medium-Sized Nuclear Reactors: You and Your Environment,* Proceedings of the 10th International Conference, San Diego, California, General Atomics, August 14-18, 1989; hereafter cited as: McDonald 1989.

60 Hahn 1988. Hahn and Nockenberg 1989a. Beyea 1989. Oeko 1989.

61 Tritium is used in nuclear weapons as a neutron source to increase explosive power. Presently the USA is considering building a new reactor to produce tritium for weapons purposes. One of the proposed choices is a MGR. If it is chosen, it will be ironic, that, yet again, the military needs will drive the development of a particular new technology.

62 Lothar Hahn and Britta Nockenberg, *Aspekte des Exports und der Militaerischen Nutzbarkeit von HTR-Modulanlagen*, Freiberg, Federal Republic of Germany, Oeko Institute e.V., Mai 1989.

63 Argonne National Laboratory (ANL), Press Release, Argonne, Illinois, ANL, November 23, 1988. Ron Winslow, "New Breeder Reactor May Operate More Safely, Produce Less Waste," *Wall Street Journal* (December 1, 1988).

64 Oeko 1989.

65 Lu 1987.

66 Oeko 1989. "Reports," *Nuclear Engineering International* (September 1989).

67 Joseph R. Egan and Shem Arungu-Olende, "Nuclear Power for the Third World?" *Anticipation*, No. 28 (December 1980). Joseph R. Egan, *Life After Death for Nuclear Power: A Global Survey of New Developments*, Report prepared for the World Commission on Environment and Development (WCED), Geneva, Switzerland, WECD, January 1986. Joseph R. Egan, *Planning a Revolution in Nuclear Power Technology*, New York, LeBoeuf, Lamb, Leiby and Macrae, 1987. Surrey 1988.

68 Stahlkopf 1989. Mcdonald 1989.

69 Radioactive wastes are also produced by hospitals, as well as research and industrial facilities, but overall in much smaller quantities than from the nuclear-fuel cycle.

70 This may be an overly optimistic assumption. From a technical point of view, disposal of high-level waste is clearly the limiting issue. Due to the much larger volume of low- and medium-level wastes, and the consequent need for larger repositories, from the point of view of public acceptance the disposal of low- and medium-level wastes may also act as constraints on nuclear growth.

71 Stephen S. Bernow and Bruce F. Biewald, "Nuclear Power Plant Decommissioning: Cost Estimation for Planning and Rate Making," *Public Utilities Fortnightly* (October 29, 1987).

72 Gordon Mackerron, "The Decontamination of Nuclear Plants," *Energy Policy* 17, No. 2 (April 1989).

73 IAEA 1989a.

74 Frank L. Parker, Robert E. Broshears, and Janos Pasztor, *The Disposal of High-Level Radioactive Waste 1984*, Volumes I and II, Stockholm, Sweden, The Beijer Institute, October 1984; hereafter cited as: Parker, Broshears, and Pasztor, 1984. Frank L. Parker, Roger E. Kasperson, Tor Leif Anderson, and Stephen A. Parker, *Technical and Sociopolitical Issues in Radioactive Waste Disposal, 1986*, Volumes I and II, Stockholm, Sweden, The Beijer Institute, May 1987; hereafter cited as: Parker *et al.* 1987. IAEA 1989b.

75 Parker *et al.* 1987.

76 Parker *et al.* 1987.

77 International Atomic Energy Agency, *Report of the Coordinated Research Programme on Environmental Evaluation and Hazard Assessment of the Separation of Actinides from Nuclear Wastes Followed by Either Transmutation or Separate Disposal*, TRS.214, Vienna, IAEA, 1976.

78 E. W. Colglazier and R. B. Langum, "Policy Conflicts in the Process of Siting Nuclear Waste Repositories," *Annual Review of Energy* 13 (1988): 317-357.

79 Parker *et al.* 1987.

80 Parker *et al.* 1987.

81 Parker, Broshears, and Pasztor, 1984. Parker *et al.* 1987. IAEA 1989b.

82 A major exception is the bilateral agreements between certain countries supplying nuclear fuel and the user countries. For example, the COMECON countries are returning spent nuclear fuel to the USSR for reprocessing and disposal there.

83 World Commission on Environment and Development, *Our Common Future*, Oxford, Oxford University Press, 1987, p. 400.

84 IAEA 1989a.

85 IAEA 1989b.

86 Oeko 1989.

87 Parker *et al.* 1987. D. Albright and H. A. Feiveson, "Plutonium Recycling and the Problem of Nuclear Proliferation," *Annual Review of Energy* 13 (1988): 239-265; hereafter cited as: Albright and Feiveson 1988.

88 Albright and Feiveson 1988.

89 United States Department of Energy, *Energy Security, A Report to the President of the United States*, Washington, D.C., United States Department of Energy,

March 1987.

[90] Roger E. Kasperson, *The Public Acceptance of Nuclear Energy: Looking to the Year 2000 and Beyond,* Paper prepared for the World Commission on Evironment and Development (WCED), Geneva, Switzerland, WCED, May 1986; hereafter cited as: Kasperson 1986. Expertengruppe Energieszenarien (EGES), *Energieszenarien: Moeglichkeiten Voraussetzungen und Konsequenzen eines Ausstiegs der Schweiz aus der Kernenergie*, Bern, Switzerland, Eidgenoessische Verkehrs- und Energie wirtschaftsdepartment, February 1988; hereafter cited as: EGES 1988. Oeko 1989.

[91] Oeko 1989.

[92] Surrey 1988.

[93] James Varley, "Three Years After Chernobyl Soviet Public Opposition Grows," *Nuclear Engineering International* (April 1989).

[94] A number of interesting studies are available, including: Roger E. Kasperson and K. David Pijawka, "Societal Response to Hazards and Major Hazard Events: Comparing Natural and Technological Hazards," *Public Administration Review*, Special Issue, No. 45 (January 1985). Kasperson 1986. Roger E. Kasperson and Jeanne X. Kasperson (editors), *Nuclear Risk Analysis in Comparative Perspective: The Impacts of Large-Scale Risk Assessment in Five Countries*, Boston, Allen and Unwin, 1987. Roger E. Kasperson, Ortwinn Renn, Paul Slovic, Halina S. Brown, Jacque Emel, Robert Goble, Jeanne X. Kasperson, and Samual Ratick, "The Social Amplification of Risk: A Conceptual Framework," *Risk Analysis* 8, No. 2 (June 1988); hereafter cited as: Kasperson *et al.* 1988. Christoph Hohenemser and Ortwin Renn, "Chernobyl's Other Legacy: Shifting Public Perceptions of Nuclear Risk," *Environment* 30, No. 3 (April 1988); hereafter cited as: Hohenemser and Renn 1988. Ortwinn Renn, "Public Responses to Chernobyl: Lessons for Risk Management and Communication," in The Uranium Institute, *Uranium and Nuclear Energy: 1987*, Proceedings of the 12th International Symposium, London, The Uranium Institute, 1988, pp. 53-56.

[95] Oeko 1989.

[96] Roger Milne, "Missing the Henkley Point," *New Scientist* 30 (September 1989).

[97] Kasperson 1986.

[98] Hohenemser and Renn 1988.

[99] Dominic Golding and Roger E. Kasperson, "Emergency Planning and Nuclear Power: Looking to the Next Accident," *Land Use Policy* 5, No. 1 (January 1988).

[100] Following the Chernobyl accident, the IAEA rapidly developed and entered into force two international conventions: Convention on Early Notification of a Nuclear Accident, and Convention on Assistance in the Case of a Nuclear Accident or Radiological Emergency. In cooperation with the FAO and the WHO, internationally agreed-upon food- intervention levels were also adopted. IAEA 1989c.

[101] For a more detailed analysis of this subject, see Oeko 1989.

[102] Christoph Hohenemser, "The Accident at Chernobyl: Health and Environmental Consequences and the Implications for Risk Management," *Annual Review of Energy* 13 (1988): 383-428. Hohenemser and Renn 1988.

[103] World Council of Churches, "Faith and Science in an Unjust World," in World Council of Churches, *Report of the World Conference on Faith, Science and the Future*, Volumes I and II, Geneva, Switzerland, World Council of Churches, 1988.

[104] For a more detailed discussion of this topic see, for example, Kasperson *et al.* 1988.

[105] While comparison of nuclear energy to alternatives is not the subject of this study, it should be mentioned that a number of recent reports indicate that, in the short and medium terms, nuclear energy is neither the fastest nor the cheapest way to provide the required energy compared to alternatives, such as increased efficiency in both the use and the supply of energy. Bill Keepin and Gregory Kats, "Greenhouse Warming: Comparative Analysis of Nuclear and Efficiency Abatement Strategies," *Energy Policy* (December 1988); hereafter cited as: Keepin and Kats 1988. David Bodansky, "Global Warming and Nuclear Power," in *Small and Medium-Sized Nuclear Reactors: You and Your Environment*, Proceedings of the 10th International Congress, San Diego, California, General Atomics, August 14-18, 1989. William Fulkerson and John E. Jones Jr., "CO_2 and the World Energy System: The Role of Nuclear Power," in *Small and Medium-Sized Reactors: You and Your Environment*, Proceedings of the 10th International Conference, San Diego, California, General Atomics, August 14-18, 1989. Oeko 1989.

[106] K. Todani, Y. M. Park, and G. H. Stevens, "The Near Term Contribution of Nuclear Energy in Reducing CO_2 Emissions in OECD Countries," in International Energy Agency (IEA) of the Organization for Economic Cooperation and Development (OECD), *Energy Technologies for Reducing Emissions of Greenhouse Gases*, Volumes I and II, Proceedings, Paris, OECD, 1989.

[107] WRI 1988.

[108] Following estimates by Edmonds and Reilly 1985.

[109] W. Gehrisch, T. Haapalainen, Y. M. Park, G. H. Stevens, and K. Todani, "The Potential Longer Term Contribution of Nuclear Energy in Reducing CO_2 Emissions in OECD Countries," in International Energy Agency (IEA) of the Organization for Economic Cooperation and Development (OECD), *Energy Technologies for Reducing Emissions of Greenhouse Gases*, Volumes I and II, Proceedings, Paris, OECD, 1989.

[110] Including the following assumptions: nuclear would always be marginally cheaper than competing coal systems; the issue of nuclear waste disposal would be resolved; no major nuclear accident would take place; and overall public acceptance would not be a concern. These are clearly very optimistic assumptions!

[111] This means a continued and increased concern over the effects of acid rain and air pollution caused by fossil fuels, but a somewhat reduced concern over the climate- change issue, following the realization that, although real, the effects would not be as serious as previously assumed.

[112] This also assumes that the nuclear generation would be replaced by coal-based electricity. However, the study did assume a 25% decrease in carbon emissions from coal plants, changes in efficiency, and increased use of cogeneration.

[113] James Varley, "COMECON Plans Still Ambitious," *Nuclear Engineering International* (June 1988).

[114] IAEA 1989a.

[115] Surrey 1988.

[116] For generic studies of this issue see, for example, Keepin and Kats 1988, and Jose Goldemberg, Thomas B. Johansson, Amulya K. Reddy, and Robert H. Williams, *Energy for a Sustainable World*, New Delhi, John Wiley and Sons, 1988. For a country study of the role of energy efficiency measures see EGES 1988.

[117] In this table we did not assume a range for carbon-dioxide emissions, and we used only one of many available projections.

SECTION THREE

Nuclear Electricity and Ethics

In the Valley of the Shadow of Three Mile Island

Joyce Hollyday
Sojourners Magazine, Washington, D.C.

"Go to the top of the hill and make a right at the Turkey Hill Minit Market; cross the railroad tracks and keep going until you pass a red brick church on your left..." As I sat in my office in Washington, D.C., taking these directions to a stranger's house over the phone, I was picturing every landmark. My sisters and I had frequently ridden our bikes over these rolling hills of central Pennsylvania. This was home.

The anticipation of returning just before Thanksgiving encouraged memories of 15 years of holiday journeys back. By this time every year, the plentiful Pennsylvania corn stalks were cleared away, leaving acres of golden stubble poking through a light dusting of snow. The geese and I sometimes met head-on, they making their way south through this valley carved out by the Susquehanna River. Seeming to point the way for them and me, white candles always glowed softly in every window of familiar, old farmhouses, whose chimneys breathed out lazy curls of smoke. Often as I hit the final stretch, a setting sun would splash brilliant red behind silhouettes of stately barns.

My parents moved to a piece of farmland in Elizabethtown about the same time I moved to Washington. They built a house that, by their description before it materialized, would be "perfect for the grandchildren," with plenty of space for Mom's flowers and Dad's vegetable garden.

I grew up down the road in Hershey - "Chocolatetown U.S.A." Hershey kids had the advantage of taking frequent tours through the chocolate factory for the reward of a free candy bar. My sisters and I spent long summer days in the town's amusement park, riding one of the oldest merry-go-rounds in the world and stopping off to sip chocolate-flavored Cokes at the counter of the Hershey drug store before heading home. I learned to water-ski on the Susquehanna at the age of 7.

We were introduced to God at the First United Methodist Church on Chocolate Avenue. For my fifth birthday, I was allowed to choose a picture of Jesus to keep. I knew immediately which one I wanted - the picture of Jesus holding the lost lamb. I kept it by my bed, believing it would comfort and protect me from all fears.

But 20 years later, a fear crept into our lives that none of us had anticipated. I was told one morning as I walked into work at the *Sojourners* magazine office, "Hershey was on the radio. They're moving pregnant women and pre-school children to the Hershey Sports Arena." I tried to picture the place where I had watched hockey games and ice shows as an evacuation center. Later that evening, the television news showed the scene: the arena lined with cots, children crying and their mothers trying to keep order under difficult circumstances. It was March 30, 1979.

The drama had begun two days earlier, when a stuck valve in the Unit Two reactor triggered a serious accident at the Three Mile Island nuclear power plant, seven miles from my parents' home. A huge, potentially explosive hydrogen bubble developed in the damaged reactor, and by day four of the crisis even some of the nuclear experts - doing their best not to provoke a panic and certainly not prone to overstatement about the danger - were beginning to talk about "meltdown" as a growing possibility.

On Sunday, April 1, my parents evacuated to my grandmother's home in Hagerstown, Maryland. The bathroom was the only truly private room in my grandmother's small apartment; I will always remember the conversation my mother and I had there. She recounted all the thoughts that had gone through her mind as they left the house, her sadness and the fear that they might never return.

She had considered bringing photograph albums and old letters along but finally decided to leave them. I had grabbed all my journals at the last minute before leaving Washington, just in case the reactor in Pennsylvania went to meltdown and forced an evacuation of the city. The fear that the future might not be the same made us both want to hang on to the joyful and poignant fragments of the past.

In the end, the hydrogen bubble mysteriously disappeared, the reactor core stopped just short of a total meltdown, and the nuclear industry assured us that no one died from TMI. The "nuclear nightmare" was over, they said. For conservative Pennsylvanians, used to nothing more dangerous in their air than the sweet smell of chocolate, the tendency was strong to believe it.

But even then, ominous predictions were offered from other sources: Cancers, leukemias, and other health effects would begin to show up in about 10 years. At the time, March 1989 seemed eons away. Four months before the 10-year anniversary of the accident, just before Thanksgiving in November 1988, I went home to find out.

A rusting sign bearing a picture of a Holstein cow and the words "Holowka Farm" swings in the wind at the end of a long, dirt lane. Marie Holowka, age 73, is making her way in an old, green pick-up truck, insistent on saving me the trouble of driving over the lane's many deep ruts, filled with water from the morning's rain.

We cannot go into the farmhouse, she explains when we reach it, because her sister is very ill. We sit in the truck, and Marie Holowka begins her story. "We were milking about 70 cows then, and we started to milk about quarter to four every morning. I think I milked about three cows, and all of a sudden the barn windows started to rattle, and the barn started to shake. And my brother said,

'Oh, my God, that's an earthquake.' And I said, 'Paul, do you hear it under the ground?' It was going brrrp, brrrp, brrrp, like boiling water in a pot. I said, 'Paul, that isn't an earthquake; that must be something happening at Three Mile Island.'" The nuclear accident at TMI began at 3:58 that morning.

Marie continues: "I finished milking about 10 minutes to seven. When I opened the door, it was just breaking day, and right away my eyes started to pinch me and there was a funny taste in my mouth. It was real blue, and I couldn't see for more than 10 feet." Marie felt so weak that morning, she fell three times on the way back to the farmhouse.

She and her brother, Paul, couldn't get any news about Three Mile Island on the local radio that morning. Finally, at 9 o'clock they heard on a Philadelphia station that there had been a major accident. The radio warned people to stay in their homes, shut all their windows, and stuff rags under their doors to keep out radioactive gases.

About 11 o'clock a county agent came and told them to feed their cows and leave them enough hay for three days. According to Marie, he said, "They should live that long if the nuclear power doesn't kill them from the accident." He also told them to pack and get ready for evacuation.

Still there was no news on local radio stations, and no word that they should evacuate. Young people who didn't have any livestock began leaving, according to Marie. Many older people stayed because they had nowhere to go. A neighbor called to say, "I just counted, and I have enough bullets to shoot my cows; I don't want them to suffer if I leave."

After the accident the Holowkas and many of their neighbors broke out in a rash that wouldn't heal. In May, Marie developed a sore throat that lasted until August, when she was diagnosed with an infected thyroid. The following August, during a routine checkup, her doctor told her she had cancer. He asked her if she felt scared. "Why should I be scared?" she said. "Everybody in our neighborhood has cancer." She believes she was hit by a radioactive plume from TMI.

Marie Holowka has had six cancer operations and 39 radiation treatments. She was taken off chemotherapy last July, because her blood is poor and the marrow in her bones has stopped rebuilding. She has lost most of her hair, her mouth is blistered, and she still has a tumor in her chest.

The week of the accident, several of the Holowkas' calves got sick, and a couple died. Since TMI first began operating, the Holowkas have lost more than 200 cows; their herd is down to 40 now. Two state veterinarians came and took blood and tissue samples from some of the dead animals. They found no diseases or poison, according to Paul.

"It used to be when a calf was born, you knew it was going to grow up," says Marie. "If it was a heifer, you knew you were going to have a cow. But not anymore. You just try to raise them, and they go so long, and then they just lay down and die."

Paul and Marie and their sister Eva have been on this farm for 52 years, since the time that their father purchased it. In their home is an array of grand-champion blue ribbons they have won for cattle they have showed in several eastern states.

They have persistently written and called Three Mile Island officials, the Nuclear Regulatory Commission, and congressional committees, trying to get an official response to their crisis. According to Paul, mostly they get told that they are making too much trouble.

Margaret Reilly of the Pennsylvania Bureau of Radiation Protection once paid a visit. Says Paul, "She tried to tell us we didn't know how to farm," alleging that the cows were overbred or underfed. According to both Paul and Marie, she brought out a machine for measuring radiation, and as soon as she turned it on, the needle shot way up. "She popped her eyes wide, and she said, 'I wonder what [TMI plant operator Metropolitan Edison] is doing today,'" says Paul. Then she told them the radioactivity was caused by uranium in the stones used to build their barn. She took a gallon of milk for testing with her when she left, but, according to Paul, he and Marie were never told the results.

Levels of radioiodine 131 were discovered in milk around the nuclear plant after the accident, and local milk companies refused to accept milk from farmers near TMI. The Holowkas now only milk for themselves and raise calves to sell. Their 163-acre farm, once thriving with livestock and fields of tomatoes, corn, and beans, has fallen into disrepair.

Paul Holowka once studied nuclear chemistry and nuclear physics at Temple University. He keeps an instrument for measuring radiation and takes readings several times a day. He has spent much of the rest of his time in the last several years taking Marie to and from the hospital.

As Marie and I jolt back out to the end of the lane, she begins, "See that big white house? Mister and Missus died of cancer in that big house - just this past year." Pointing to a stone farmhouse, she continues, "And right there, he was 41, he died about three years ago from cancer. Right across the road there, she died right after the accident; she was 31 years old. There the man had the rash so bad, he finally died with it. There was a little boy here died in the hospital - his heart was on top of his chest. Not last summer, but the summer before that, we had 14 neighbors died right around here. This is practically a new neighborhood."

Marie and Paul Holowka, along with many of their neighbors, are convinced that their troubles stem from the towers visible from the edge of their property - the two that belch steam around the clock as well as the two that now stand ominously silent. They can't prove that their conviction is true. And no one can disprove it.

They, and others like them, are at the heart of a decade-long debate that has drawn in doctors and lawyers, mothers and legislators, pastors and farmers and some of the best minds in the history of the nuclear industry. What everyone seems to agree on, from the operators of the Three Mile Island plant to the psychologists who have done documented studies, is that the trauma of the accident has raised levels of stress, depression, and fear among people in the vicinity of the plant.

But there the concensus seems to end. What remains are personal stories and, among those for whom Three Mile Island is more than just a bad memory, a heated controversy that boils, it seems, just short of meltdown.

In May of 1979, Mary Osborn's daughter carried a bouquet of daisies home to her after school. Among the bunch were two severely deformed flowers, and when her daughter led her to the field where she had picked them, Mary found about 25 more.

In October Mary Osborn sent both of her young children out into the yard to collect leaves for crayon rubbings. The maple leaves they brought back were too large to fit under the paper. Seven months before, on the morning of the accident, she recalls the air seeming "full of metal--you couldn't tell if you were swallowing it or breathing it."

Mary Osborn is not a botanist or radiation specialist. She is a mother who simply began taking note of an array of flower and plant deformities in her neighborhood, and then in other communities around the Three Mile Island plant.

As sun streams through her picture window, past hanging ferns and impatiens and begonias, she brings out a large, flat box wrapped in gold ribbon. Carefully untying it, she pulls out huge maple leaves pressed between sheets of newspaper and a now-brittle, 31-inch dandelion leaf picked on a ridge near TMI. One of the maple leaves, she says, repeatedly registered radioactivity when a Geiger counter was passed over it. In small boxes she has a two-headed miniature sunflower, fused and leathery maple leaves, dandelions with thickened and curled stems, and a rosebud on a long stem growing from the center of another bloom.

Osborn was invited to West Germany to talk about Three Mile Island with people in the path of the radioactive cloud that left Chernobyl after the Soviet reactor's disastrous accident in April 1986. She noted that people had found particular plant abnormalities in the same progression that she had. Some of the people she met gave her specimens of the strange flowers that had come up in their yards and gardens. One man talked about finding some unusual dandelions - with leaves more than 30 inches long.

Mary Osborn has a signed affadavit, presented in a hearing before the Nuclear Regulatory Commission, from world-renowned botanist Dr. James E. Gunckel, who has researched the effects of radiation on plant growth and development for 39 years at the Brookhaven National Laboratory and Rutgers University. Detailing types of plant effects and radiation, his affadavit states that, after carefully examining plant specimens after the TMI accident, he found "a number of anomalies entirely comparable to those induced by ionizing radiation."

Both of Osborn's children have been hospitalized at times, she says, that correlate with announced releases of radioactivity from TMI - her daughter for a week with a fever of unknown origin and her son with respiratory trouble. He went in on a New Year's Day. Mary says that at the hospital "there were babies in the aisles - there were no more cribs for kids." She is quick to caution that not every ailment should be blamed on the power plant, but adds that "when you can correlate something that happened down there with what's happened to you or your family, it really gets bizarre."

If anything, Jane Lee is outspoken and forthright, a constant thorn in the side of the nuclear industry and the public officials from whom she has demanded a response regarding problems in the wake of Three Mile Island. She is also a constant giver of care to the animals that live in the shadow of the tall, aqua silos that distinguish this farm from the others in the rural town of Etters, on the west bank of the Susquehanna.

Lee painstakingly tells the story of a duck, born unable to walk, flopping around behind its mother. When the duck lost her down, the feathers that came in, instead of being smooth and flat, were brittle and stuck up like quills. Lee fashioned a sling on a long string for under her body, and tried to "do some therapy." One person kidded her, "You know, the neighbors see you walking a duck, you're in trouble."

The duck lived only five months. She was perhaps lucky simply to have been alive at all. After the TMI accident, according to Lee, great numbers of eggs never hatched.

Lee began to notice some problems around the farm - in operation since the first German settlers came here 200 years ago - even before the accident at Three Mile Island, not long after the nuclear power plant began operating. She is very concerned about the "routine" releases of radioactivity from the plant, as well as the consequences of the accident.

She began to see an increase in spontaneous abortions and stillbirths on the farm. Some of the calves that were born alive were unable to walk. She mentioned these problems to Dr. Robert Weber, a veterinarian who has had a practice in the area for 35 years. "You're not the only one," she says he replied. "We're having problems all over the place." When Lee asked him what he meant, he said, "I don't know what's going on, but in a five-mile radius of that plant down there, we've got serious problems on the farms all around," she recounts.

Weber mentioned that he normally performed one or two caesarean deliveries a year, but after the accident he was doing about one a week, and he said recently to Lee, "Now I'm going day and night." The animals are suffering from constriction of the cervix and are simply not able to deliver their young.

With some research, Lee discovered that the reproductive systems of animals are particularly vulnerable to low-level radiation. One of the affected hormones, relaxin, is the one that triggers and prepares the body for labor. The Three Mile Island accident happened at the worst possible time for animals, she points out - the females were just beginning to give birth to their young.

Litters of kittens came out stillborn and hairless; dogs were born without eyes; other animals were born dwarfed or retarded. Some livestock developed raw sores and lesions, others muscle deficiencies. Pigs had trouble breeding, as did the

Dauphin County coroner's five throroughbred horses, according to Lee. She has carefully documented a range of problems among the animals in the vicinity of the plant.

Jane Lee is disturbed about the dismissal of the problems by public officials. She pulls out a slim report issued by the Nuclear Regulatory Commission and the Environmental Protection Agency in October 1980. "Did you ever see a government report this thin?" she asks. The report concludes that none of the reported plant and animal problems experienced around TMI could be linked to the power plant. "They came in here and talked to us and then they went back and wrote this thing up," she says. "And they did it in a matter of days." There were no soil or water or feed samples examined, and no blood or tissue samples taken from animals, according to Lee.

"The animals are giving us clear signals as to what's wrong in our environment," says Lee. "Human beings don't seem to understand that what happens to the animals who eat and sleep and feed in the environment is also going to happen to humans."

Just as she was starting to notice widespread birthing problems among animals, she read an article about a coroner expressing concern about a surge in crib deaths in York County. She went to a library and sorted through newspaper obituaries, finding a substantial increase in stillbirths and crib deaths in the area she believed had received a heavy-dose plume of radiation.

Lee later uncovered a report by Dr. Alice Stewart, one of the world's most respected radiological epidemiologists, credited with discovering the link between x-rays and fetal abnormalities. In this paper, Dr. Stewart detailed how the exposure of fetuses to radiation depletes the oxygen in their blood. After birth, the breathing of these children becomes sublevel, and they are unable to push away the mucus in their breathing passages. "There are a lot of women walking around today," says Lee, "blaming themselves because their infant suffocated."

Lee brings out a large map of the TMI area, with a multitude of overlays. Some, she says, show the paths of radioactive plumes from the nuclear power plant. Others show health effects - cancers and cancer deaths, thyroid problems, respiratory ailments, leukemias, skin rashes, and reports of "metallic taste." One solitary dot marks a dentist's office where 75 unexposed films in a desk drawer were discovered exposed after the accident - "there's a definite correlation between that and gamma radiation," says Lee. Jane Lee is persistent in pressing for answers from the government and the nuclear industry. "If I keep quiet, then I'm as bad as they are for letting it happen," she explains.

I ask what she thinks it will take to get the industry to respond. She lays out a frightening scenario: "There is going to be an accident so bad, with breach of containment, that they will not be able to cover up the effects. It will have to happen in a place like New York City or Chicago where a lot of people die. If they want the bodies stacked up, then maybe they'll listen to what we were trying to tell them. When that happens, it will be the end of nuclear power. It won't happen before that."

Joyce Corradi sits at the low table where the children she keeps for daycare every day eat lunch and do artwork. She is surrounded by pine boughs and ribbon and works on a Christmas wreath as we talk in her Middletown home. Her five children are in an adjoining room singing along with a record playing "O Come, All Ye Faithful." Her youngest, Mark, interrupts to ask her to cut the wheels off of a small toy skateboard he wants to transform into a surfboard.

Her life, Corradi says, was her family and the local PTA. That was before TMI. It is clear as she talks that concern for her family is still at the heart of what she does - she has just taken that concern into new and unlikely arenas.

Corradi, some neighbors, and women in her Catholic parish set out to do what they felt the government and nuclear industry had "grossly failed" to do in the wake of the accident - safeguard the health of their children. They formed Concerned Mothers and Women, an organization that, as Corradi describes it, is made up of women "about as average as you can come across."

Acknowledging that they had more concern than expertise, the women began meeting regularly with nuclear industry and political officials, as well as members of the Nuclear Regulatory Commission, demanding answers and accountability. The attitude of Metropolitan Edison toward them at first, she says, was, "You are basically a nuisance; please go home and bake your cookies."

Corradi was disturbed by the arrogance of the experts. "Those people were lying," she says. "I mean, first there was no accident; then there was a little accident; and then there might be a big accident but we really know how to handle it," was the word from the industry. She adds, "Now [scientists] are bringing out more and more data that say the accident was indeed very grave," referring to recently released reports that the temperature inside the reactor exceeded 5,000 degrees and 50 percent of the reactor core melted.

Eventually, she says, it got "to the point where [the nuclear industry] had told so many lies to so many people that the facade was beginning to crack" and they had to respond to groups like Concerned Mothers and Women. The best thing the group has done, she adds, is to "say to many people, in many areas, women as mothers and citizens have very strong concerns for the health and safety of their children, and that cannot be ignored."

The experience of the Three Mile Island accident has changed her life, according to Corradi. For one thing, she thinks she and her friends should be awarded Ph.Ds for all the technical knowledge they have acquired over the past decade. Before TMI, she would never have pictured herself going in her Sunday clothes after church to picket at the gates of a nuclear power plant, where one afternoon she encountered a tourist with TMI on his list of attractions ("abominable," she says) who labeled her a communist.

Joyce Corradi is outspoken and confrontative of the powers that be, but she also harbors a gentle spirit of reconciliation. Several years ago, when TMI clean-up operations after the accident reached the dangerous point of lifting the head off the damaged reactor, the media wanted a comment from Concerned Mothers and Women. Corradi told them that she and all the other mothers were taking their children to a state park 40 miles away for the day. "But," she added, "our prayers and good wishes are with the men who are physically going to go down there and do that dangerous work" on behalf of the community. She told representatives of the nuclear company that concern for the workers was "a bridge that we both can cross, no matter how I feel about you people. "

She recently attacked deficiencies in evacuation strategies for future crises. She points out that, during the last evacuation, most of Harrisburg's black population - "anyone who was poor or disabled" - didn't get out. She says that last time around, people in Middletown were naive about the dangers, and adds, "The next time there is an evacuation, people are going to kill to get out of this community."

Joyce Corradi acknowledges that she is like "David against Goliath," and many people have asked her what good she is doing when the nuclear industry and government always get their way. She answers, "I can't look at my five kids and say, `They were going to do what they wanted anyway, so why should I bother?' I have to say to them, `I tried.'"

The reason she keeps on is rooted in what happened to her oldest son, Tony, then 9, at the time that the family evacuated. With strong emotion and disturbingly graphic detail, she describes the scene: "My son threw up a vile green slime, the color of Comet cleanser, and it scared the hell out of me. And to this day, other than two doctors from Japan who told me that it was a reaction to radiation, no one has ever confirmed what that was or what that may have done to my son."

The presence of Japanese doctors in the area after the accident was both disturbing and comforting. They had done studies on survivors of the Hiroshima and Nagasaki atomic bombings and came expecting to find more data on radiation victims. When Joyce told them about her son, one pulled out a pile of paint tiles and asked her to point to the color of what he had expelled. The seriousness with which they took the problem was in direct contrast to government officials who told Joyce it "must have been something he ate."

When Corradi conceived Mark after the accident, she was very concerned about his health. When he was born, he had growths on his ears that had to be surgically removed. The pediatric surgeon could not explain them. "Was it just a hereditary fluke, or was it something that was caused by the accident?" asks Joyce. "Who can answer?

"I believe that there are some real medical problems that we may have to face down the road," she continues. "To tell my children otherwise is to do what I have accused the [nuclear power] company of doing. If I said to them, `There's no problem,' that would be like saying, `Let's not tell them the truth; let's pretend we have an answer.'"

Corradi talks again about her oldest son. "He turned 19 yesterday," she says. "He seems healthy, he's productive, he's happy and, please God, let it be that way. But who's to say that somewhere down the road it won't be? What happens when he has children, or when his children have children? Those are questions children should not have to live with."

The lack of response from the officials who should have been collecting health data moved some citizens to take action. Norman and Marjorie Aamodt, who had years of technical experience conducting surveys for Bell Laboratories, developed a methodology for a health survey. Dr. Carl Johnson, a physician and public health official in Colorado who documented an increase of cancers around the Rocky Flats nuclear weapons plant, developed a questionnaire, and physician Bruce Molholt of the Environmental Protection Agency lent his support. Several citizens, including Mary Osborn, Jane Lee, and Joyce Corradi, went door to door interviewing in April and May of 1984. In the three areas they surveyed, they discovered a sevenfold increase over the expected number of cancers.

The Aamodts sent their report to the Nuclear Regulatory Commission to receive verification. The NRC forwarded it to the Centers for Disease Control. However, Marjorie says, the NRC enlarged the pages so that the numbers were lost and only sent about half the report, which was dismissed by the CDC. Independent scientists later verified the Aamodts' methodology and conclusions.

The report was disturbing, and the Pennsylvania Department of Health immediately moved to undermine it, accusing the Aamodts in a press conference of faulty methodology and, worse, of lying, according to Marjorie. The Department of Health carried out its own cancer study, released in the fall of 1985, which concluded that no adverse health effects had been found in people living in prescribed 5- and 10-mile radius zones from Three Mile Island.

The Department of Health study was discredited by both Dr. Robert A. Hultquist, a Pennsylvania State University professor of statistics, and Dr. George Hutchison, a Harvard University professor of epidemiology, according to a report in the *Harrisburg Patriot-News*. They concluded that the facts had been diluted by the inclusion of 150,000 people who lived beyond the designated study zones - more than 40 percent of those factored in the study.

Despite the Department of Health's attempted assurances, at least 2,000 people in the area believe that they have experienced health effects as a result of the TMI accident - strongly enough to have brought lawsuits against Metropolitan Edison and GPU Nuclear Corporation (General Public Utilities, the reorganized offshoot of Metropolitan Edison that took over management of the plant after the accident). The lawsuits represent, as one reporter put it, "an enormous catalog of human suffering"- cancers of all types, miscarriages, birth defects, leukemias, kidney and thyroid problems.

Some of the suits have already been settled out of court, although details have not been made public; it is widely believed that part of the settlement of these cases is that the victims are not to speak about them. Cases involving children by law must be made public, and several million dollars have been paid out to their families. A representative of GPU stated clearly that these settlements do not admit guilt on the part of the company; they were simply settled to avoid court costs.

The remaining cases have moved between federal and state courts, and the overwhelming number has led a Pennsylvania judge to order plaintiffs and defendants to pick six cases each, to be tried as representatives of all the rest. Fred Speaker, attorney for the defendants, says that a key argument for the defense is that the statute of limitations has run out on many of the cases.

According to Speaker, studies show that persons in the vicinity of the plant received radiation the equivalent of "less than one chest x-ray." He says the burden of proof rests with the plaintiffs, adding, "Unless we win all of these cases, there will be claims filed probably for the next 30 years." Asked his chances, he says, "I think we have a very good shot at this, ultimately. We've won almost all of the legal battles so far." Arnold Levin, attorney for the plaintiffs, calls the actions of Metropolitan Edison/GPU at the time of the accident "grossly negligent....They withheld information from the public and made statements saying there was nothing to worry about. If that's not irresponsible conduct, then I don't know what is."

"What's at stake," says Eric Epstein of the citizens group Three Mile Island Alert, "is the entire nuclear power industry. If the health claims are proven out, the whole industry would be indicted."

The contention all along of the Pennsylvania Department of Health, as well as the nuclear industry, has been that not enough radiation was released from the TMI accident to have created the health problems being reported. Margaret Reilly of the Bureau of Radiation Protection said that the amount of radioactivity released at the time of the accident amounted to a "gnat's eyelash."

The official information put out by GPU Nuclear Corporation lists common health complaints and offers condescending dismissals of them: Metallic taste may be a result of dietary supplements; irritation of the eyes can be caused by infections, allergies, and chemical irritants; rectal bleeding is a common symptom of hemorrhoids; skin rashes result from cosmetics, exposure to sunlight, and "normal aging processes."

One of the most disturbing realities about the accident is that no one will ever know precisely how much radiation was released. The Three Mile Island Public Health Fund, which was established after the early legal settlements, commissioned a preliminary study of radiation monitoring at the time of the accident. The report was released in August 1984 by principal investigator Dr. Jan Beyea, senior energy scientist with the National Audubon Society. The study found that "the monitoring network in place, both inside and outside the plant, did not perform adequately"; that "environmental sampling...was insufficiently coordinated, with problems in labeling and calibration"; and that "a great deal of crucial data does not exist, or is unreliable."

The study reported that the thermoluminescent dosimeters (TLDs), used to detect environmental radiation, appeared to be spaced too far apart, leaving gaps in monitoring. The calibration of the filters used for in-plant measurements of released radioiodine was questioned, and filters were discovered missing after the accident. The study refers to reports that suggest that "many hundreds of times more radioiodine was released... than was estimated...in the official studies." It mentions the possibility that large amounts of radioiodine escaped unmonitored into the ground or the Susquehanna River.

Dr. Beyea's November 1988 summation of the situation at the time of the accident was this: "There wasn't enough radiation equipment; half of it wasn't working; and a large amount went off scale, because they [nuclear industry officials] never anticipated such large releases."

He says the Aamodt health study is "a good exploratory study that indicates an excess level of cancer" and commends the Aamodts' "good technical backgrounds" and careful methodology.

He questions whether the radiation released was high enough by itself to create such a dramatic rise in cancers and believes that work needs to be done to determine whether chemicals released from the plant, some perhaps carcinogenic, are a factor.

Beyea concludes with a reference to current efforts to discover the truth about the effects of the accident: "It's an extremely important issue. If one ever wanted to find an example of how not to deal with public concerns, TMI is the prime example."

A December 1987 report on radiation monitoring at TMI, released by the Three Mile Island Public Health Fund, concluded that, almost nine years after the accident, significant gaps still existed in monitoring practices. A January 1988 study released by the Institute for Energy and Environmental Research (IEER), a two-year effort by a multi-disciplinary team of eight scientists headed by IEER Executive Director Dr. Bernd Franke, is more direct. It labels current monitoring practices at Three Mile Island "highly inadequate." It continues, "The emergency monitoring system is so deficient that if an accident occurs during unfavorable weather conditions, the emergency response decision could be fatally wrong."

Based on the findings of the study, Franke concludes, "It appears that environmental monitoring is more of a public relations exercise rather than an effort to design and build a system which will reliably detect emissions once they occur."

He is working to reconstruct dose levels from the accident. "People were exposed to higher doses than acknowledged. There were 'hot spots' in the area," he says, referring to places affected by plumes of high-level radioactivity. These were not detected, he says, "simply because the plume does not always do you the favor of landing on a detection instrument."

Thomas Gerusky, director of the Bureau of Radiation Protection, says, "I do not believe that there was enough radioactivity released to cause any health effects. I think the only health effects we're seeing are psychological."

When asked about gaps in the data and problems with monitoring, he replies, "No, that's not really true. You can't monitor every square foot of every spot on the land. However, the monitoring that was there was more than adequate to cover the accident, in my opinion." He "doesn't buy" the theory that there were gaps in the monitoring that might have missed plumes of high radioactivity. He says that during the accident, aircraft and people on the ground were monitoring constantly, "and we did not see the kind of levels that would indicate that there was a significant, abnormal release that we didn't know about."

Margaret Reilly of the same bureau says, "I don't know that I would characterize what was there during the accident as inadequate; it was basically comparable with what everybody else had at the time." She says that the Nuclear Regulatory Commission, GPU, and the bureau have all increased their TLDs around the plant. When asked about the missing data, she says, "There was an inhouse air sample filter that was missing, which upset a lot of people. I have no idea what happened to it....But it would have filled an interesting hole, as it were."

Reilly disputes the Holowkas' version of her visit to their farm. "I had a civil defense Geiger counter, which is not the best thing in the world for that sort of measurement. And even if I were to turn on one of those in the office today, you would see the needle bounce back and forth....We have better stuff for looking at really low levels [of radiation] now than we did."

She does say that she might have told the Holowkas that there was uranium in the stones of their barn. "But it should not have been detectable," she says. "It's not enough to see legitimate changes with a Geiger counter."

She continues, "The thing is, in this business, it is so rare that one finds actual radioactivity that when you find it, as a technocrat, one tends to jump on it....We have all this equipment, and we have people sitting around looking for almost nothing, and if we ever find it, it's looked on as something interesting to get into. And if we think we have something, we certainly follow it."

Twelve days before the TMI accident, Dr. Gordon MacLeod, a friend of then Pennsylvania Governor Richard Thornburgh, assumed his post as the state's secretary of health. During the accident, MacLeod pressed for more medical input on the emergency and urgently called for an evacuation of pregnant women and young children. The crisis was two days old before Thornburgh gave the notice to evacuate.

MacLeod also tried fervently to secure from the federal government a supply of potassium iodide, a liquid that can be taken to block the ingestion of radioiodine and thereby protect the thyroid from damage. The supply arrived five days after the accident - too late for it to do any good - and thousands of the vials were unlabeled or contaminated.

During the inquiry by the Kemeny Commission - the presidential commission established by Jimmy Carter to investigate the TMI accident - Thomas Gerusky made what MacLeod calls "misstatements." MacLeod later wrote a letter to members of the commission concerning these statements.

In October 1979, seven months after MacLeod assumed his post, Governor Thornburgh asked for his resignation. MacLeod says, "The only reason given for my requested resignation was a difference in operational style. Well, the governor asked me to serve as secretary of health knowing my operational style - and it didn't change."

When asked about MacLeod's firing, Thomas Gerusky says, "You'd have to ask Governor Thornburgh; I wasn't involved." Thornburgh was unavailable for comment. As to the incident of MacLeod's letter to the Kemeny Commission, Gerusky says, "I don't recall that."

In some sense, Dr. Gordon MacLeod may have been prophetic at the time of the accident. Simply by consulting the state's vital statistics for the months following the crisis, he found dramatic increases in both infant mortality and hypothyroidism in the areas surrounding TMI. Dr. George Tokuhata, director of the Division of Epidemiology Research of the Pennsylvania Department of Health, claimed the rise in infant mortality was rooted in the high concentration of minorities - particularly low-income blacks - in Harrisburg. Says MacLeod, "That high concentration of minorities existed before the accident - but this was reported as if it was a valid argument."

Shortly after MacLeod left as secretary of health and returned to his professorship at the University of Pittsburgh School of Medicine, someone still in the Pennsylvania Department of Health alerted him to the high rise of hypothyroid cases among children. It was critical that the data be released, says MacLeod, so that the condition could be treated and cretinism avoided in the children. The data were "sitting on someone's desk for three months," he says, until MacLeod himself released the information to the press.

MacLeod's continuing concern is that decisions about nuclear power are being made without adequate health input. Despite his raising the issue 10 years ago, there are still no physicians on the staff of the Nuclear Regulatory Commission, according to MacLeod, and so the future health of people is left in the hands of "a group of physicists, engineers, technicians, and administrators."

Since 1952, Dr. Ernest Sternglass has worked on detecting and measuring radiation. He has developed technology to reduce x-ray doses and was a pioneer in the development of the imaging technique of fluoroscopy. He is currently professor emeritus of radiological physics at the University of Pittsburgh School of Medicine.

He arrived in Harrisburg the day after the TMI disaster began. "I was very concerned," Sternglass says, "because I had taken a Geiger counter with me, and as we were flying into the airport, the readings went up between five and ten times the normal background [radiation]. Riding in the taxicab into the city, I noticed the meter going up and down, indicating large pockets of radioactive air....I knew that whether or not there was going to be a meltdown that would lead to a catastrophic release...certainly the most vulnerable individuals - young children and women who were pregnant - should be immediately evacuated."

Sternglass talks of the false assumption that has fueled the nuclear industry for decades: Extrapolating from data on survivors of the Hiroshima and Nagasaki bombings, it was concluded that low doses of radiation have very little effect. But, says Sternglass, there is not a direct, linear relationship between dose and effect, and low-level radiation is much more harmful than has generally been accepted.

Sternglass says that "there's no question" that Gerusky, Tokuhata, and others intentionally distorted and suppressed health data. He says, "In fact, there is just no other way to explain the changes in the vital statistics that took place the following year."

He found that the U.S. vital statistics, published monthly, showed a peak of infant mortality during the first year after the accident. "The following year," he says, "these peaks disappeared from the revised data." He also discovered that infant mortality dropped 20 to 25 percent during the time that the plant was shut down. "A year or two later," says Sternglass, "the data were revised so as to fill in this valley, thereby elimating the evidence in the final printed volume of the U.S. vital statistics." He also uncovered that data showing a rise in deaths at the time of the accident were deliberately withheld until after the presidential commission had made its report.

Sternglass says we have "only seen the tip of the iceberg. Many of the cancers take 10 to 15 to 20 years to show up. All we can say is that we have greatly increased the chances of premature death and congenital defects and loss of immune resistance in our population around...nuclear reactors." He quotes a study on mortality in the United States by Dr. Jay Gould of Public Data Access, who "found a clear rise after 1979 amounting to a minimum of 50,000 and as high as 90,000" deaths in one year.

During the week that I was in the area of Three Mile Island, the Pennsylvania Department of Health released another health study, which concluded that there was no evidence of increased cancer in the area. The only effect it found was low birth weight among some infants, which was the result of mothers taking extra medication, such as tranquilizers and sleeping pills, at the time of the accident, according to Dr. George Tokuhata.

When I mention that a previous Department of Health study was soundly discredited by Drs. Hutchison and Hultquist, Tokuhata says, "I was not aware of their comments." When I quote the source of the information, he says, "I remember reading newspaper articles - in local newspapers." But, he adds, "There was no conclusion there - just a critique."

When I bring up Dr. Sternglass' allegations, Tokuhata responds, "He's been saying this all over the place - Savannah River, Colorado [Rocky Flats]" - referring to Sternglass' work on health problems in the vicinity of nuclear weapons production plants. "There's absolutely no truth in it - he's that kind of a person....He has evidence absolutely no one would agree with."

But, in fact, there are many people in the area surrounding Three Mile Island who are convinced of its dangers and potential dangers. After the disaster in TMI's Unit Two reactor, Unit One was shut down. When NRC hearings were initiated to determine GPU's competence to run the plant, citizens mobilized to try to prevent the restart of Unit One.

The record of Metropolitan Edison and GPU was not what one would call clean. Documented records showed that Unit Two, which had been in operation for only three months before the accident, had a history of maintenance-caused failures in its main water feed system, which was the source of the accident.

The official U.S. House of Representatives committee report on the accident, issued March 1981, offered this indictment: "...TMI managers did not communicate information in their possession that they understood to be related to the severity of the situation....[which] prevented State and Federal officials from accurately assessing the condition of the plant. In addition, the record indicates that TMI managers presented State and Federal officials misleading statements (i.e...inaccurate and incomplete) that conveyed the impression the accident was substantially less severe and the situation more under control than what the managers themselves believed and what was in fact the case."

Also in 1981 the NRC held hearings on a cheating scandal that was uncovered among TMI's operators in training. Even more scandalous, it was discovered that the power company had been falsifying data on radioactive release rates within the plant. According to Marjorie Aamodt, while researching NRC documents she found evidence that the NRC was aware of the falsification but had done nothing about it.

In 1983 GPU Nuclear Corporation was indicted on 11 criminal counts related to falsifying and destroying safety data. The corporation pleaded guilty and was convicted on two of the counts; it pleaded no contest to the other nine.

It came to light later, as reported in *The Toronto Star,* that the late, prominent Naval nuclear expert Admiral Hyman G. Rickover - on whose staff a few of GPU's officers had served - had written reports supporting restart of Unit One in return for a donation of $380,000 by GPU to an educational foundation he had formed, according to his son, Robert Rickover. Further, Rickover's daughter-in-law, Jane Rickover, swore in a notorized affadavit that her father-in-law had told her that he had used his personal influence with then President Jimmy Carter to pressure him to "suppress the most alarming respects" of the presidential commission report on the TMI accident and release it in "a highly diluted form." Admiral Rickover had told her, according to the affadavit, that "the report, if published in its entirety, would have destroyed the civilian nuclear power industry, because the accident at Three Mile Island was infinitely more dangerous than was ever made public."

In January 1985, alleging that radioactive releases from TMI had created health and plant effects, that Metropolitan Edison/GPU had tried to deceive the public about the seriousness of the accident, and that the data on radioactivity that the company claims were lost during the accident "were more likely intentionally destroyed to prevent disclosure of the hazard," the Aamodts brought a motion before the Nuclear Regulatory Commission to conduct further hearings before licensing the company to restart Unit One. Their motion was denied.

The NRC set the restart date for June 11, 1985. In response, the Aamodts, the Union of Concerned Scientists, Three Mile Island Alert, and even Governor Richard Thornburgh (who later became President Ronald Reagan's attorney general) joined in a legal appeal to block the restart on the grounds that GPU's character and competence to run the plant safely had not yet been proven. The Aamodts eventually took their plea all the way to the Supreme Court.

On May 18, 1982, citizens in three counties surrounding the plant voted 2-to-1 in a non-binding referendum to keep the plant shut down. Concerned citizens, including Eric Epstein and Mary Osborn, repeatedly participated in actions of nonviolent civil disobedience, blocking TMI's gates on several occasions. In one case, the jury found the defendants guilty, but on the basis of the expert testimony offered in the trial, issued a statement calling for keeping Unit One shut down.

In a May 1984 rally organized by Concerned Mothers and Women, a Pennsylvania state representative said, "We are here to say no. We have said no. How many more times will we have to say it?" Joel Roth, vice chair of an NRC Advisory Panel, said, according to *The Harrisburg Patriot-News*, "I'd like to see a big sign...that says, 'This nuclear facility was closed because it was the will of the people.'"

Despite all the citizen opposition, the criminal conviction of GPU, and increasing complaints of health problems, on October 9, 1985, TMI's Unit One reactor was put back in operation. Eric Epstein says, "What became clear was that the plant was going to restart no matter what we did. As a result, probably one of the greatest tolls...is the demise of democracy and...[people's] faith in democracy."

He says of GPU, "They've brought nothing but heartaches and bad times to this community. They have done so many...incredibly devastating things to this community, there's no way they can regain the trust. When you have an accident with a core that exceeds 5,000 degrees, and you put people through the nightmare that you put them through, it's very hard to go out in a three-piece suit and talk nice and convince them that everything is OK, especially when you're a convicted felon."

Gordon Tomb, media relations manager for GPU Nuclear, meets me at the bridge that leads over to Three Mile Island. As we drive through the plant complex, on streets with names like Liberty and Justice, he points to the huge casks of radioactive waste that have been taken from the damaged reactor and are on their way by rail to the Idaho National Engineering Laboratory, run by the Department of Energy (DOE).

Inside a control room, we watch on video screen the operations inside the reactor. Workers dressed in bulky protective clothing, in constant radio contact with the control room, are attempting what has never been done before - digging through a damaged reactor core and mounds of highly radioactive debris to get to the bottom of the accident at TMI. Some of the work has had to be done underwater, and the clean-up process had just been delayed by six months due to problems cutting through steel plating around the fuel.

The future of TMI's Unit Two reactor is another point of hot contention in central Pennsylvania. GPU wants to put it in "Post Defueling Monitored Storage" - or "mothballing," as it is more commonly called, for at least 30 years. According to Eric Epstein, GPU is "postponing [the decontamination and decommissioning], hoping that the technology will be developed." Meanwhile, he says, that leaves Three Mile Island "an unstable, radioactive waste site" well into the next century.

Gordon Tomb admits that during the accident the company gave out "conflicting information, confusing information, and inadequate information," attributing it to the company being "ill-prepared" for the kind of accident it faced. He also expresses regret that the accident "frightened people."

But he says that the "amount of radiation released has been established," and it was not enough to have created the kind of health effects people in the area have been reporting. He guesses that most of these people "are completely sincere in what they say and believe," but adds, "Honestly, sincerity isn't good enough." He says that as far as health concerns, "It's been pretty much put to rest - maybe that's an overstatement, because there are 2,000-plus cases pending -but as far as the industry is concerned, it's been put to rest."

He acknowledges that GPU pleaded guilty to two indictments, but adds, "I'm not sure that [records] were actually falsified - but it was at least sloppy." Of other violations that have been documented by Three Mile Island Alert, some involving issuing of warnings or fines by the NRC - including plant operators asleep on the job, use of drugs and alcohol by operators, incidents of radioactive contamination in the plant, and improper safety procedures - Tomb says the incidents "typically have some basis in fact, but are typically skewed to seek TMI Alert's purposes." He adds that a violation read out of context "sounds, quite frankly, a lot worse than it is." He points out that in the year ending March 1988, TMI's Unit One reactor had the world's fifth-best operating record on the basis of capacity.

Of the referendum to keep Unit One shut down, Tomb says, "These issues are too complicated and not appropriately addressed in a referendum," then adds, "We have an obligation to our customers to provide electricity." Picking up the theme later, he says, "We need clean, safe energy for the well-being of this country and the world. There are a lot of hungry people out there whose well-being depends on economical power as a real important source."

Stored in huge tanks on Three Mile Island is a major by-product of the accident that will not easily go away - 2.3 million gallons of radioactive water.

What should happen to it is as heated a controversy as the debate over health effects.

GPU's first plan was to dump the radioactive water into the Susquehanna River, according to Frances Skolnick, who is an "intervenor" on behalf of citizens in the TMI area concerned about the water. Citizens successfully blocked that plan. The current GPU proposal is to evaporate the water and release the radioactive gas into the air. Quoting a friend, Skolnick says of GPU's alternatives, "They're asking, 'Do we want to be shot in the foot, or do we want to be shot in the arm?'"

Skolnick and others are particularly concerned about the evaporation plan in light of the highly toxic, radioactive tritium that would be released to the environment. Says Skolnick, "It's a totally unnecessary exposure to low-level radiation." She continues, "People have been onslaughted by the accident and then 10 years of continuous exposure....Why do they want to subject people to any more?"

Skolnick feels that the nuclear industry wants to set a precedent. If the evaporation is successful, it will be seen as the "perfect cure" for liquid radioactive waste. The resolution of the issue has major implications for the nuclear industry. As a representative of citizens in the area, Skolnick is advocating that the water remain stored on the island until its radioactivity decays.

I met Frances Skolnick at hearings by the Atomic Safety and Licensing Board (ASLB), where she faced a phalanx of lawyers representing the NRC and GPU, who were armed with reams of testimony by physicians and radiation experts on the safety of the evaporation plan. She explains that the ASLB judges are paid by the same body that pays the NRC lawyers and that, rather than getting the truth on the record as is their mandate, "As every day passed, it was obvious that their objective was to undermine me."

Frances Skolnick has pursued the crusade on behalf of central Pennsylvania's citizens at great personal cost. "I won't quit," she says, "because I'm speaking for my two children, who are not at the point of their lives where they're able to stand up and speak. And I'm speaking for my friends' children, too. We're deciding the future for them."

Tomb says that the amount of radiation that would be released by evaporating the water "isn't even measurable."
He dismisses the opinions of well-known nuclear scientists Dr. Karl Morgan, who testified before the Atomic Safety and Licensing Board on the evaporation issue, and Dr. John Gofman. He says that their opinions "represent the views of an extreme minority of scientists" and "are not considered credible by the experts in the field." He says that "reputable scientists" concur that the radiation releases at the time of the accident were too small to cause health effects, and that they "had nothing to lose but their own credibility and reputations if they sought to distort the records."

Dr. John Gofman tells a different story. Gofman is professor emeritus of medical physics at the University of California at Berkeley. He has received worldwide acclaim for his work on heart disease. He is a pioneer in radiation research who co-discovered the fissionability of Uranium-233 and isolated the world's first workable quantities of plutonium for the Manhattan Project, the enterprise that developed the atomic bomb, during the 1940s.

In 1963 Gofman became associate director of the Lawrence Livermore Radiation Laboratory and founder of its Biomedical Research Division. He and a colleague were given the task to find out the health effects of radiation. They came to the conclusion that radiation was 20 times more harmful in producing cancer than was thought. Gofman released this information publicly.

"Up to that moment," he says, "the AEC [Atomic Energy Commission, forerunnner of the Nuclear Regulatory Commission] considered me a `fair-haired boy,' because I didn't say anything that they didn't want to have said." But, he says, at that moment, "the roof caved in on me. And they started essentially a war against...me." They took away all Gofman's laboratory research funds. Before his study, one AEC commissioner had told him, "All we want you to do is tell the truth," according to Gofman, but, he says, "There was nothing they wanted less than the truth."

"I don't have any respect for the NRC," he says. "I think they are much more concerned with the financial health of the utilities...and they tend always to minimize health effects of radiation. So I don't trust anything that comes out under the auspices of the NRC...or the DOE. My experience with them is that they're no better than the old AEC. They tend to hide things."

Indeed the entire sweep of nuclear history - from the dropping of the bomb on Hiroshima, to bomb tests in the Pacific Islands and the Nevada Test Site, to the current controversy over high incidences of cancer in the vicinity of nuclear weapons production plants - is an immense catalog of attempts to cover up the effects of radiation and marginalize and discredit those scientists who, like Gofman, have attempted to expose the truth. This abysmal history is well-documented in *Killing Our Own*, by Harvey Wasserman and Norman Solomon.

In that book is an astounding quote by John Gofman. Gofman was part of the team that developed the bomb, the initial rationale for which was stopping Nazi atrocities and experimentation on humans. But, Gofman says, in the United States "we have already accepted the policy of experimentation on involuntary human subjects....I feel that at least several hundred scientists trained in the biomedical aspect of atomic energy - myself definitely included - are candidates for Nuremberg-type trials for crimes against humanity through our gross negligence and irresponsibility." And, he adds, "Now that we know the hazard of low-dose radiation, the crime is not experimentation - it's murder." Gofman stands by this quote, adding, "I have repeatedly stated that setting permitted levels of radiation exposure for people is the same as saying you think it's permissible to have a certain amount of random murders."

Gofman says that the Three Mile Island accident remains a "kind of mystery," its health effects uncertain. He adds that it was "just a lucky break for all of us" that the reactor didn't go on to a complete meltdown.

Gofman says, "I don't believe for a minute that they [the nuclear industry] really believe in the safety of their reactors." He mentions the Price-Anderson Act, which essentially absolves the industry of liability for its accidents. "If you won't put your money where your mouth is, you don't believe what you're saying--it's just that obvious." He wants to see the nuclear power industry shut down, he says, because, "I can only foresee...that they're going to steadily contaminate the earth more, and there will be a very high price the people will pay for it."

Ten years after the accident that shook central Pennsylvania and the nation, the fallout continues. For many, the accident was a life-changing experience that still carries deep spiritual and emotional impact.

Rev. Gregory Harbaugh received his formal call to serve as associate pastor of the Lakeside Lutheran Church in Harrisburg the weekend TMI blew. His first response was, "Am I going to have a place to go to?" He found a congregation that was dealing with a great deal of stress, confusion, and fear.

Harbaugh sees nuclear power as "not only a dangerous way to boil water, but also a dangerous way of concentrating power in the hands of a relatively few people." In the aftermath of the accident, his congregation became divided over TMI, and some pro-nuclear members left the church. He says, "People are still anxious about it, but I think they largely feel powerless to do anything."

After the accident, people often asked Harbaugh, "Where is God in all this?" He says, "I was saying, 'God is right here.'... Most of us have an image of God as Superman - 'Why didn't Superman come in and rescue us from this terrible thing?' We tend to divorce God from the cross, so that the concept of God present with us, suffering with us, doesn't register very well."

Joyce Corradi remembers that after the accident, the bishop of her Catholic diocese came to her parish and granted the entire church general absolution, or pardon for all sins. She says, "That is unheard of, except in a very dire crisis" such as extreme emergencies or times of war.

Some people talk about the converting power of the accident. David Messner, an administrator in the Pennsylvania Department of Community Affairs, says he used to be "a preacher of left-wing politics" who had "absorbed the doomsday scenarios by environmentalists uncritically." The accident pushed him to read about nuclear power, and he became convinced that the media were exaggerating the danger.

After the accident, Messner and others formed Friends and Family of TMI, which sells souvenirs at the TMI visitors center and has promoted nuclear power through such efforts as providing prizes for pro-nuclear science projects at area schools. Messner feels that the risk to TMI workers "substantially increases when they leave the island and drive out on Route 441" and believes that nuclear power "gives the Third World an option out of chopping down the rainforests to generate electricity."

But it appears that the overwhelming number of conversions has been in the other direction. Rep. Bruce Smith, a Pennsylvania state legislator, was chair of the board of Newberry Township at the time of the accident. He says, "The news media called me for my reaction to the accident, but I hadn't been informed of it....The state police went to neighborhoods [to alert people to evacuate], but they skipped some, including mine."

Smith, whose views represent well those of his people in this largely Republican, conservative area whose major newspaper is called *The Patriot-News*, says, "I'm a different person....I'm more skeptical than I used to be. The experience of TMI made me much more sympathetic to the plight of the individual who feels the system is against him or her."

He continues, "Prior to the accident, I had no problems with nuclear power whatsoever. After the experience at TMI, I firmly believe that nuclear power plants should not be built near populated areas. I think most Americans feel that way."

Powerlessness, feelings of betrayal and the demise of public trust, fear of the future and one's children's future, continuing stress on marrages and families - all of these are part of the fallout of Three Mile Island. In the midst of all the controversy over health effects, it must be remembered that the psychological consequences of the accident are documented health effects that are widespread and real.

Rev. Earle Fike Jr., who was pastor of the Elizabethtown Church of the Brethren, says of the accident, "Those who lived through it walk around with a dread in their soul." Indeed, the fear in this valley is tangible. And the memories are vivid.

"I was in class, in second grade, and I remember that my Aunt Margie and the principal came and told me to come out. My aunt seemed kind of worried.

"My aunt brought me home, and my Mom was getting everything ready to go. That was when I heard that there was an accident at the power plant, and it was dangerous. The kids were telling stories - like that somebody's grandma was sitting in her rocking chair and all of a sudden she died from this thing. And I believed it; I didn't know what to expect. But I wasn't really concerned about it then, because I was little; and when you're little, you think your parents will take care of everything and won't let anything happen to you."

The old clock on the mantle chimes out quarter-hours as Beth Longenecker speaks, a reminder of the passage of time, as her story brings home the fact that 10 years is time enough for a little girl to grow into a young woman anticipating college. Indeed, this whole house speaks of time, of rootedness and generations.

Beth's parents both have roots in this area that go back 200 years. Like many people here, their ancestry is *Deutsch*, or German, which generations ago got mistranslated as "Pennyslvania Dutch." I share these roots on my mother's side, and I grew up familiar with a tradition well-known for colorful quilts and decorated barns, for hospitality and generosity and plentiful good food.

The good food is evident here, as Patricia Longenecker, Beth's mother, insists on feeding me a meal that is traditionally Pennsylvania Dutch New Year's fare -

pork and saurkraut, mashed potatoes and homegrown vegetables, homemade applesauce, and, of course, a generous slab of shoofly pie.

Patricia's great-grandfather was the last canal boat operator on the Susquehanna River. Other relatives once had a watermelon farm on Three Mile Island, which, she says, "grew the best watermelon in the area because it was on the flood plain." After a major flood in 1929, they sold the island to the power company.

Sixteen years ago, not long after Beth was born, Patricia and her husband, John, cleared land and built this house outside Elizabethtown. It looks like a homestead out of *Country Living*, with care given to every detail - baskets filled with dried flowers, quilted wall hangings, and rustic hurricane lamps. In the upstairs hall is an old wooden chest, built in 1832 of poplar and pine; Patricia's son will be the eighth generation to inherit it.

Patricia picks up the story of the TMI accident where Beth left off, recalling the instant when she realized just how serious it was. Her voice breaks as she says, "At that moment, time stopped."

She describes the scene in the town: "Everyone was in a state of panic, and no one knew what to do. It was fueled by the sirens, and by broadcasts through the town. People just began leaving quickly, some not even knowing where they would go....By Sunday Elizabethtown had become a ghost town."

She talks about the conversion she experienced: "We went from total belief in a system--feeling secure that the [nuclear power] corporation, and certainly the federal government, would only promote something for everyone's welfare - to total shock and disbelief that this would be allowed to happen." Patricia began devoting her time to setting up citizens hearings regarding the TMI accident and promoting alternative energy sources, particularly solar.

Both Patricia and Beth talk about meeting Dr. John Gofman and the tremendous impact of that encounter. When Beth realized that he had been "fired because the government wanted him to say something he couldn't," she lost her trust in the system: "I feel that the government is hiding so much that they're not telling us." When I ask her how that makes her feel, she says, "Scared."

Two years ago, when she was a sophomore, Beth delivered a "persuasive speech" in class against TMI, an opinion that is not shared by most of her peers, many of whom have parents who work at the plant. She explains, "When you're a teen-ager, you have more on your mind than finding out what the government is doing."

At a young age, Beth carries an awesome and sometimes lonely sense of responsibilty for the truth. She is headed to college to study sports medicine, or perhaps environmental science. Her essay on her college application is an articulate plea for care of the environment. "We have to start doing something," she says. "It might even be too late."

Patricia speaks of a spiritual pledge that is this home's foundation: "When we made a commitment here 16 years ago to create a homestead from the woods, we looked at this whole project with a caretaker's spirit--to grow our own food, to try to be as self-sufficient as possible, to give back to the land what we took." She mentions the deer who come to the salt lick, the ducks who gather at the lake, the abundant fruits and vegetables that stock their dinner table. "You must give the care in order to have something to care for. That's what life is all about. You give, and you always receive - and more so, tenfold.

"And this is the message on the larger scale," she continues. "If we're not taking care of the environment, protecting it...what are we going to have then to care for?"

The Longenecker homestead is in Lancaster County, Pennsylvania, considered among the most fertile farmland in the world. "A nuclear plant came a half hour to meltdown - which could have meant the end of a whole farming region," Patricia says. "Here we are 10 years later - what have we learned?" She adds that she has been "bombarded by friends having cancer."

Beth then talks about a woman who lived on a ridge near their home, the mother of three young children, who just found out she has cancer. I ask her if she has fear for her own health and future. "Yes, I do," she says. Her mother says Beth is convinced she will have cancer some day.

I leave the Longeneckers' lane and turn left - toward the clouds of steam rising from the horizon. It is a portentous night, accented with a full moon shining between patchy clouds. A fierce wind howls through bare trees, sending dead leaves airborne in circles from the ground.

As I approach them, the cooling towers rise tall - red lights blinking, pink steam swirling all around like vapors from some boiling, primordial soup. The view is frightening, as if coming face to face with the spewing mouth of an ancient abyss of unknown proportion and danger. Ezekial's apocalyptic vision of Gog and Magog, the enemies from the north, comes to mind.

Behind the belching twin beasts sit two silent partners, eerily quiet. I shudder and turn from the view. "We would all be better off with watermelons," I say to myself as I head toward the other end of Elizabethtown.

During the time that I was home, another of my parents' friends died of cancer. He was a neighbor for several years, the father of my first childhood friend.

I do not know what gave him cancer. I only know that disturbing things are happening in the Susquehanna Valley - and in other places in this country in the shadow of nuclear power and weapons plants.

Upon hearing the news of his death, I thought back to the day in Sunday school 30 years ago when I chose the picture of Jesus holding the lost lamb. It has been a comforting image for a long time. In moments of crisis or fear, I have often turned to the 23rd psalm, to that vision of a shepherd who leads and protects and consoles.

In all my readings and prayings of that psalm, the images of pastures and still waters were always vivid; but I lacked a clear picture of the "valley of the shadow of death." Today, that image is disturbingly concrete for me. It is home.

Ethics of the Nuclear Option in the 1990s

Rosalie Bertell
International Institute of Concern for Public Health
Toronto, Canada

Although the nuclear-energy debate has raged within the churches since the 1970s, some of the basic moral questions have yet to be dealt with adequately. Having been one of the ethicists on the National Council of Churches Task Force in the United States (1978-79), and involved in the preparation for the WCC MIT meeting, I am in a position to say this. More information is now available, and I believe that the moral stance is clearer today than in 1980. Some of the problems which we then saw, as independent scientists, are now openly admitted by the nuclear industry. Important recent happenings include the re-analysis of the atomic bomb data;[1] the publication in January, 1990 of the *Health Effects of Exposure to Ionizing Radiation*, by the US National Academy of Science;[2] and the recommendation of more stringent limits for radiation exposures of nuclear workers and the general public by the International Commission on Radiological Protection (ICRP) in February 1990.[3] The cost of the past mistakes of the nuclear industry has fallen disproportionately on the world's indigenous people. They have borne the burden of uranium mining, milling, and nuclear-waste disposal generally.

In this chapter, I would like to put forth for your consideration some basic ethical statements that I shall consider in some detail in the main text:

- Nuclear fission, in itself, is morally neutral and, with proper safeguards, may be used to produce electricity.
- Currently used nuclear technology is ethically unacceptable because its design basis permits what we now know to be unacceptable levels of ionizing-radiation exposures to workers and to the general public. The engineering design needs to be drastically changed. Transfer of these outmoded design systems to developing countries is unethical.
- The serious threat to the gene pool posed by radiation has never been dealt with, ethically, by the nuclear industries or standard-setting agencies. This jeopardizes the viability of future generations and places the major risks incurred for our benefits on our descendants.
- A proper method of documentation for worker health and for damage to the health of the general public, or to the environment, has not been implemented by governments committed to nuclear technology, by the nuclear industry, or by the International Atomic Energy Agency (IAEA). This failure to document harm cannot be taken as proof that harm has not occurred.

- By current international standards, the nuclear industry, as it now operates, is not safe. Workers should be receiving hazard pay; general radiation-protection standards should be improved, and the general public and food supply given more protection. Mechanisms for base-line community health studies and for documentation of worker health problems are needed. This action needs to be accompanied by stronger safety precautions, both within and outside the facility.
- No solution has been found over the last 45 years for the safe handling and isolation of nuclear waste. Hence many real costs are being charged to "sparsely populated areas" and to future generations. As radiation-protection standards are tightened, economic problems will increase. Ocean dumping or contamination of economically developing countries is unacceptable. The combined effect of damage to the gene pool and pollution of the air, water, and land is intolerable.
- Both the normal operation and the possibility of catastrophic accidents make current nuclear technology, including the breeder reactor, unacceptable. The range of lethality possible with nuclear reactors exceeds that of all other energy sources.

Just as the true costs of fossil-fuel usage, including black-lung disease, acid rain, and carbon-dioxide pollution (Greenhouse Effect), required a century to become manifest, so too we have been slow to identify the real problems of current nuclear technology. Perhaps the nuclear-energy problem has been overshadowed by dialogue on the nuclear-bomb issues, the toxicity of plutonium, and risk-benefit equations which sometimes obscure, rather than clarify, the difficulties. A hard look at the nuclear generation of electricity reveals deep ethical problems beyond those traditionally dealt with by the churches.

1. Nuclear Energy Is Not Morally Neutral

This principle has had its history of debate in the churches. Since electricity can be produced by an eel, an automobile engine, or a battery, using uranium fuel to produce electricity is certainly at the "more dangerous" end of the range of production sources. Nuclear reactors basically boil water, fuelling steam generators, hence are a source of electricity. Nuclear energy cannot be used to replace coal, oil, or gas as fuel for automobiles, planes, or ships. Neither can it replace oil in producing plastics or medicines. It is, then, a possible source of heat for the production of electricity. As part of an ethical inquiry, I believe the first question to be addressed is whether electricity is needed for end uses that include space heating, plastics manufacturing, and supplying liquid fuels. If it is, then what source of electricity is "best" for the people and the planet?

While it is possible to use electricity to power air conditioners, for example, it is also possible to use direct solar energy to cool buildings or warm water, as has been demonstrated in Darwin, Australia.[4] Need for more electricity is a real moral question which requires an answer before the nuclear option is considered.[5]

2. Current Nuclear Technology Has a Flawed Foundation

Historically, the recommended levels of radiation exposures for workers and members of the general public were developed between 1945 and 1950 by the Tolerance Dose Panel, a group of British, Canadian, and American physicists who had worked on the atomic bombs. These permissible exposure levels, 5 rem per year for workers (or 50 mSv) and 0.5 rem per year for the general public (or 5 mSv), remained the recommendations of the International Commission on Radiological Protection (ICRP) until February 1990.[6] The current recommendations would lower worker exposures to below 2 rem (20 mSv) per year and general public exposures to below 0.1 rem (1 mSv) per year.[7] Based on current risk estimates of cancer deaths, these recommendations are still too high.[8]

It was in the late 50s and 60s that the commercial nuclear industry developed. The nuclear engineers accepted the radiation-exposure recommendations as a "given" and tried to develop an industry within these guidelines which would be cost-competitive with coal. They were often very conscientious members of the churches who strongly defended the nuclear industry on the basis that its pollution was well within the maximum permissible levels. Unfortunately, they assumed "legal" was equivalent with "safe." We now know that, even when the average radiation exposure of workers is maintained at 20% of the maximum permissible level (1 rem or 10 mSv per year), and the average exposure of the general public is maintained at 1% of the maximum permissible level (5 mrem or 0.05 mSv), the radiation-induced cancer level is unacceptable by the usual standards for rating hazardous industries. (This problem is discussed more thoroughly in section five of this chapter.)

The radiobiological mistake made in 1950 by the Tolerance Dose Panel haunts the nuclear industry today. Lowering permissible exposure levels means re-designing the workplace and the gaseous and liquid radioactive release systems. Admitting the mistake and returning to the drawing board is now a very expensive process. Ignoring the mistake, at the cost of more lives, is morally reprehensible. Britain, Japan, and the US have already begun to adjust to the change. The British National Radiological Protection Board recommended lowering permissible exposure of workers to 1.5 rem (15 mSv) a year. Similar groups in Japan and the US recommend a limit of 2 rem (20 mSv). In the US, the public exposure from a local nuclear facility must be less than 0.025 rem (0.25 mSv).

How will these stricter exposure standards be incorporated in sales of reactors to developing countries? Will they receive the outmoded technology, as happened to the Philippines? The Philippine reactor was not upgraded to safety requirements which became mandatory in the US after the Three-Mile-Island accident, because these were not mandatory at the time the deal was struck.

The draft recommendations of ICRP released in February 1990 are vulnerable to further reductions. For example, the ICRP has assumed that the risk of cancer death is constant, regardless of a population's historical cancer rate. This assumption has been rejected in favor of a relative risk model by the US National

Academy of Sciences.[9] The relative-risk assumption increases the number of malignant tumors by a factor of three, and leukemias by a factor of four. Moreover, the ICRP divides the cancer risk by two when extrapolating from the high-dose, fast-dose rate to effects at the low-dose, slow-dose rate. In contrast, using the revised 1986 dosimetry for the atomic-bomb survivors and mortality rates through 1985, one finds 15 excess cancer deaths per 100 person-Sieverts at high doses of radiation, and 35 excess cancer deaths per 100 person-Sieverts at low doses of radiation.[10,11] The rate most likely needs to be increased with chronic exposure, not arbitrarily reduced by 50%. Presumably, at higher doses there is widespread tissue death and lifeshortening from other causes. Moreover, as the biological endpoint moves from fatal cancers to genetic effects in offspring, even more drastic cuts in permissible exposure levels are expected. These changes in regulatory limits must be provided for in any new reactor-purchase contract.

Even without using the new, higher-radiation-risk levels, there are millions of casualties due to nuclear activities since 1945.[12] The new risk estimates mean past damage to health has been much greater than people realized.[13]

3. Radiation Degrades the Gene Pool

The "health effect of concern" with radiation exposure has changed over the years. In 1928, the effect of concern was radiation burns. In 1950, it was changed to concern for initiating fatal cancers. In the 1930s, people had found that cancers were produced in the women who painted radium on watch dials to make them glow in the dark. The levels of radiation received by the women were below those which caused burns. Hence the change to a new biological endpoint brought about a lowering of permissible exposure levels.

There is also evidence that the Dose Tolerance Committee in the 1940s was well aware that genetic damage had been caused by radiation that was well below the levels recommended as "permissible," based on cancer induction. A paper by D. G. Catchside was submitted to the committee for discussion on February 6, 1947 with a plea that they remember the devastating effect of radiation on the human sperm and ovum when setting permissible radiation exposures. A note by Dr. D. E. Lea was appended to the document. Lea, a well respected radiation expert and a member of the Dose Tolerance Committee stated (memo dated April 28, 1948):

> There will be no means of showing whether the abnormalities in the child (of an exposed worker) were due to the occupational exposure of the parent, and so no question of liability for compensation can occur....
>
> It will not be possible to prove whether the effect in any particular instance (of hereditary semi-sterility) was caused by radiation, so no question of liability for compensation can occur....
>
> so long as less than 1% of the population is exposed to radiation there is not likely to be a noticeable increase in the incidence of hereditary abnormalities.

Put rather crudely, the argument of Lea for not dealing with genetic damage, with basic justice to workers and their children, is the inability of the victims to prove damages. He argues that this inability assures that the industry will not be held liable for compensation, and therefore the damage can be ignored.

Catchside pressed his argument further, stating that the gradual undermining of health of children would eventually lead to noticeable reduction in their vigor. K. Mather, Department of Genetics, University of Birmingham, UK, a Tolerance Dose Committee member, accepted the general thesis of a slow-but-sure "rise in the loss of fitness" in the population with chronic exposure to increased levels of ionizing radiation. He took courage in the "fact" that no more than one percent of the population would be exposed. (This was 1950, before widespread nuclear testing and reactor expansion.) He also notes that:

> In particular, in human society, the fate of an individual is not independent of the activity of his fellows...If his fellows take steps to alleviate his lot, his selective disadvantage will, at least in some cases, be reduced and his contribution to the next generation increased....The point to be made clear is that, in the case of man, discussion of the effect on the community of raising the mutation rate cannot be divorced from consideration of changes of selective forces, in the way that is possible with other species.[14]

He suggests that the "solution" might be to eliminate the less fit. Today amniocentesis, with abortion of the most defective offspring, fulfills this role.

Two well-known members of the Tolerance Dose Panel went on to become members of the International Commission of Radiological Protection (ICRP), namely Sir E. Rock-Carling and Professor Mayneord, but the question of genetic effects received no more attention until 1959. In 1959 the ICRP adopted the linear hypothesis, that is, the assumption that radiation damage will be proportional to doses even below the maximum permissible dose level. Prior to 1959, many assumed that radiation-induced cancer and genetic damage, like burns, had a threshold or safe level below which there would be no harm. The threshold myth can still be heard today. However, it is now generally agreed that neither effect has a threshold, i.e., even the lowest doses of ionizing radiation can damage the genetic material within the nucleus of a cell.

In 1957 Herman Muller again introduced the genetic question to the standard recommending organization, the International Commission on Radiation Protection (ICRP).[15] His view was clearly stated in his article, "Radiation and Heredity," in the *American Journal of Public Health*.[16] He predicted the gradual loss of vigor as more generations were exposed to ionizing radiation. My own research led me to the same conclusion, independently, in 1977 as a result of a ten-year study of 16 million people exposed to ordinary diagnostic medical X-rays.[17] Persons who start life with genetic defects have reduced ability to adjust to stress or to cope with the infections and environmental hazards of life. Their ability to survive is undermined.

The First International Congress of Human Genetics, meeting in Copenhagen August 7-11, 1956, submitted the following resolution to the World Health Organization:

> The damage produced by ionizing radiation on the hereditary material is real and should be taken seriously into consideration in both the peaceful and military uses of nuclear energy as well as in all medical, commercial, and industrial practices in which X-rays or other ionizing radiation is emitted. It is recommended that the investigation of the amount and type of damage and of related genetic questions, be greatly extended and intensified with a view to safe-guarding the well-being of future generations.[18]

The World Health Organization Study Group on Human Heredity also, in 1957, called for serious investigations of human populations before countries made further commitment to nuclear industries.[19] However, even the January 1990 BEIR V Report states: "The effects of radiation on the genes and chromosomes of reproductive cells are well characterized in the mouse."[20] The committee has continued to use mouse data to estimate genetic effects of radiation. Studies of the genetic effects of the population of Kerala, India, exposed to high background radiation for generations, had never been undertaken by governments. With the help of WCC and a team of Indian physicians and scientists, our Institute has been gathering data on congenital diseases in Kerala. We hope soon to release the findings. This population at risk had been clearly identified for study by the WHO in its 1957 report.

The neglect of genetic data in favor of cancers appears related to the composition of the standard recommending body, the ICRP. Between 1950 and 1985, there have been 45 commissioners of ICRP, each serving an average of 12.6 years (4- to 24-year range) In this time period, three commissioners were geneticists, serving 34 person-years or 6.0% of the total time commitment. There were one radiobiologist and one biophysicist who served as commissioners. Membership has been as follows:

physicists:	16 men (45.5% of person years)
medical administrators:	11 men (25.8% of person years)
radiologists:	12 men (14.9% of person years)

These three disciplines represent 87% of the members and 86.2% of the person-years of service on the ICRP. They have been drawn from government agencies and from nuclear-weapons programs in a self-perpetuating manner. Members of the ICRP propose and accept new members in an internal process. They invite outsiders to serve on sub-committees, but these have no real power to make recommendations. Only commissioners have votes. No outside body such as the WHO can recommend or place members on the ICRP.[21]

Geneticists, radiobiologists, epidemiologists, physicians trained in occupational or public health, or other medical experts (one would assume would be involved in setting radiation exposure standards) have either never been invited to be commissioners of ICRP, or were in a small minority position on the commission. Although the problem of accumulation of genetic mutations in a population has been admitted to be serious, no standards have addressed it, and no monitoring has been devised to measure it.

4. Documentation of Worker, Genetic, and Community Health

It has been well known, since the 1940s, that it would require new methods of data collection to document changes in worker health, or the even more subtle damage to offspring, to the community, or to the environment from ionizing radiation.[22] In spite of this, there has never been a properly set-up study, initiated by either government or the industry itself to document these and estimate the magnitude of effects. A few death-certificate studies of children around nuclear facilities and of workers have reported serious problems.[23,24] The death certificate telescopes one's entire medical history into "first cause of death," neglecting years of disability with non-fatal chronic disease. Children's death certificates contain neither parent's employment information nor even the child's place of birth and birth weight. Neither data source gives length of residence in an area. There have recently also been some studies of first cause of death, initiated by public pressure or even by TV stations, but there is still resistance to research and a minimizing of the findings. Since the ICRP minutes of the 1950s show that the commission understood that mutagenic and carcinogenic radiation damage would be difficult to prove, and since we now have evidence that inadequate record keeping continues to make proof difficult for the victims, we can conclude that the nuclear nations intended to maintain the "no-liability-possible" status quo. The public-relations pitch of the industry states: "there is no record of the death of a member of the public from commercial nuclear reactors." Such a claim becomes even more cruel and unethical in the light of the history of the ICRP and its ignoring radiation hazards.

Even the prestigious *Handbook of Radiation Management and Protection* states: "Some people may become sincerely concerned about nuclear-reactor safety despite the perfect safety record." The context of this remark is a section in the *Handbook* on the Three-Mile-Island accident, in which the author admits that it was the worst accident in the history of US nuclear-power reactors. "Still," he says, "no one was killed or even injured."[25] There is no mention of the case, now before the US Supreme Court, which documented 20 cancer deaths in the TMI downwind area where 2.8 would have been expected, and 11 out of 19 pregnancies terminating in neonatal death or congenital anomalies.[26] The government studied only psychosomatic illness in the vicinity of the plant. Record-keeping on human health has been consistently called for by such prestigious organizations as the World Health Organization,[27] and the lack of record keeping has been decried by investigative committees.[28] Yet, few records have been kept to document health effects of radiation.

It is certainly reprehensible for an industry to claim safety on the basis of non-studies while admitting the difficulty, or even impossibility, of the victim's proving liability. Publishing definitive statements of "no harm," in an international reference book, immediately after an accident, when the radiation releases are known to act over a long time period, is unconscionable.

On two occasions I have been able to show health damage related to normal operation of commercial nuclear reactors,[29] and to exposure to uranium-radium waste within levels claimed to be safe by the nuclear industry.[30] These studies were done with little or no funding assistance other than charitable donations. In both cases, my findings were ignored, or critiques were secretly distributed, so that I had no way to reply to the so-called "scientific dissent." The second paper was reviewed by three nuclear engineers who declared it not good epidemiology, a judgment totally outside their academic competence. Moreover, they based their conclusion on a two-page summary, rather than on the professional paper. This is obviously not ethical, or even scientific, behavior on the part of the nuclear industry and government.

My advice to economically developing countries at this time is not to accept health and safety claims of the nuclear industry. To 1985, a conservative global cumulative estimate of cancers in the population attributable to civilian nuclear reactors and their support industries is between 2,700 and 8,000, based on United-Nations' estimates of exposure.[31] By the year 2000, the number is expected to rise from 10,000 to 31,800. About half will be the direct cause of death. Genetic diseases to 1985 are between 2,100 and 100,000 (depending on whether or not one includes diseases with a genetic component). By the year 2000, the number will be between 8,400 and 420,000. The estimate of cancers in nuclear and uranium workers was 3,240 to 9,720 by 1985, and 12,800 to 38,400 by the year 2000.[32] It is tragic that only a few workers and no members of the public have thus far been able to hold the nuclear industry liable for this damage. In Canada the Workers Compensation Board has recognized 170 uranium miners' lung cancers as radiation-related, and about 200 miners are still awaiting action. In addition, Czechoslovakian studies claim that uranium-miner deaths are really occurring at a rate of three to four times higher than is admitted in North America.[33,34]

In developing countries, many of the radiation-exposed people will die of infectious diseases caused by a depressed immune system prior to clinical diagnosis of cancer. These radiation-related deaths will probably never be attributed to the nuclear industry, since it will take years to develop mutually acceptable liability criteria, even for diagnosed cancers. By the time a community understands the true health costs, the reactor will be at the end of its useful lifetime, and its financial assets will be usurped with decommissioning and waste-disposal problems.

5. Nuclear Generators Are Not "Safe"

A "safe" industry is usually defined as one causing no more than one worker death per 10,000 workers per year, and causing no more than one death in the general public per million persons per year.[35]

There has been considerable disagreement, over the years, as to the number of fatal cancers initiated by a measured dose of ionizing radiation. The recent re-analysis of the atomic bomb data (1987) has assigned new (lower) doses to the Japanese survivors, bringing the cancer estimates, per dose, higher and more in line with those proposed by independent scientists over the past 20 years. The nuclear industry now needs to adjust its permissible worker-radiation exposures to lower levels, in order to claim "safe workplace" status. The progression of estimates, including those of the United Nations' Scientific Committee on the Effects of Atomic Radiation are given in Table I. (Each estimate assumes average worker exposure is 20% of the maximum permissible exposure.) It is quite clear that the two official estimates, from the UNSCEAR and ICRP publications, are very low. They are the source of pro-nuclear public-relations material. These agencies are expected to be forced to increase their estimates because of international scientific pressures within the next year. In February 1990, the first draft for lowering worker levels of radiation exposure was released by the ICRP.[36]

The BEIR estimates (Table I) are from the US National Academy of Sciences. This estimate was updated in 1990 by BEIR V to reflect the new dosimetry at Hiroshima and Nagasaki. My estimate will soon move to 15 to 20 deaths since, as yet, I have not fully incorporated the new Hiroshima data. My findings incorporated the atomic bomb estimates reported prior to 1985.[37] There is also some debate in the economically developed countries about whether or not cancer deaths should be the biological endpoint counted, especially with life-prolonging treatments. Frequently a person with cancer dies of pneumonia, heart disease, or some other cause.

The Hanford Study (Table I), directly based on the death certificates of nuclear workers, yields an estimate of 26 deaths per 10,000 workers per year. This study was done by Dr. Thomas Mancuso, University of Pittsburgh Medical School, with Dr. Alice Stewart and George Kneale of the Queen's Tumour Registry, Birmingham, England.[38] The industry has claimed that some of these deaths were due to workplace hazards that are not radioactive. For our purposes, this is irrelevant. There may well be other substances in the workplace that initiate fatal cancers in the workers, especially since the safety of an industry is judged by the number of worker deaths, per 10,000 workers per year, regardless of which workplace hazard causes death. A "safe" industry has a yearly average of one death.

Dr. John Gofman was former Director of Radiation Health at the US Lawrence Livermore Nuclear Laboratory. He opposed US nuclear-weapons testing in the 1960s. His estimates (Table I) are based on his analysis of the published atomic-bomb data. His credentials as a medical doctor, as well as a (Ph.D.) physicist, place him in a unique position to evaluate the information. My estimate was based on a review of the published (peer reviewed) scientific literature.[39] The number (Table I) represents a central figure of the range of reported estimates as of 1985.

Rudy Nussbaum, a (Ph.D.) physicist from Portland, Oregon, studied the atomic-bomb survivor data only for lower-dose exposure (below 100 rad or 1 Sievert). About 80% of the survivors are in this subgroup. His estimates (Table I) are made directly from the data and are not extrapolated from the highest-dose groups.[40] As you can see by the last figure, the linear hypothesis, the low-dose direct-estimate based on Hanford workers and lower-exposure atomic-bomb survivors is higher than one gets from the linear hypothesis. The nuclear industry has consistently claimed that the linear hypothesis is overly conservative and overestimates the cancer rate at low doses. This is not true, probably because, at higher doses, more cells are killed. Dead cells cannot cause cancer.

D. L. Preston is responsible for the official updated atomic-bomb-survivor reports from the Radiation Effects Research Foundation in Hiroshima. The reason for his estimate (Table I), 17 deaths, being slightly lower than Dr. Nussbaum's, is that Nussbaum corrected for 23% under-reporting of cancer in survivors of the atomic bomb; he also eliminated the persons with doses over 100 rem (1 Sv). The under-reporting factor is not contested, but official estimates do not incorporate it. As will be noted, Preston's estimate of 17 is higher than the linear dose estimate, 11 deaths.

Since 1987, the US atomic-bomb-based estimates have moved closer to the estimates of independent scientists. The A-bomb researchers have admitted that the doses assigned to survivors in 1965 were incorrect. Another source of atomic-bomb error was the inclusion of persons with very high doses in the cancer estimates. This error has not yet been corrected. High doses kill cells or render them sterile, reducing the number of cancers per unit dose. Low doses are much more relevant for estimating cancers in workers or in the general public from nuclear industries. As time goes on, more atomic-bomb survivors die and the risk estimates are altered. Those under ten years of age in 1945 were 55 years old in 1990, an age of increased cancer vulnerability.

Nuclear workers are clearly in a very hazardous industry compared to other industries which handle cancer-causing substances. (See Table II.) According to the UNSCEAR 1982 Report to the United Nations, occupational exposures for nuclear workers in the US achieved the average levels given in Table III. These data represent a consistent, successful effort to keep the average worker-exposure level below 20% of the maximum permissible level of exposure. In order to achieve "safe industry" status, one death per 10,000 workers per year, given the data which indicates that 1 rad (10 mSv) per year would initiate 17 to 41 fatal cancers, the nuclear industry would have to achieve new average exposure levels at least below 0.06 rem or 0.6 mSv per year. This is a drastic reduction in exposure level to 1.2% of the permissible level of 5 rem or 50 mSv per year for workers. The new worker average to be achieved is 12% of the maximum permissible dose now in use for children, pregnant women, the elderly, and other members of the general public. That level is also obviously too high at this time.

The scaling down of worker exposures, including those of uranium miners and millers, fuel processors, reactor operators, reprocessing workers, transportation workers, and waste handlers requires a major expensive redesigning of the industry. The corresponding reduction for exposure of members of the general public requires further major design questions, even in the United States where worker exposures are relatively low. It also has implications for the design of nuclear-waste repositories.

Other countries likewise will need changes in exposure levels and therefore working conditions.[41]

Canada:	Annual average radiation exposures reported in nuclear reactors during 1977 varied from 0.08 to 0.91 rem (0.8 to 9.1 mSv).
Argentina:	Annual average radiation exposures for 1977-79 varied from 1.4 to 1.7 rem (14 to 17 mSv).
United Kingdom:	Annual average radiation exposure at nuclear reactors for 1975-79 varied from 0.22 to 0.27 rem (2.2 to 2.7 mSv).

Generally the mechanical-maintenance workers at a nuclear reactor receive the highest average doses annually; in Canada (1979), this is 1.15 rem or 11.5 mSv (millisieverts) per year. Worker doses at the Eurochemic Fuel Reprocessing Plant at Mol, Belgium, 1970-78, ranged from 0.9 to 1.8 rem (9 to 18 mSv) per year. All of these nuclear industries must be drastically redesigned to reach a "safe-industry" exposure level below 0.06 rem (0.6 mSv) per year.

Actually, as in any major industry, the nuclear utilities experience industrial accidents and worker exposure to other hazardous materials. The radiation exposures may well need to be further reduced to allow for non-radiation-related deaths and to maintain an average of one worker death per 10,000 workers per year.

The accepted criterion for a safe industry, from the point of view of the general public at risk from its activities, is that it cause no more than one death per million people per year. The current maximum permissible radiation dose to the public is 500 mrem or 5 mSv, and the ICRP proposed reduction is to 100 mrem or 1 mSv per year.[42] In the US, public exposures from a local nuclear facility must be below 25 mrem or 0.25 mSv per year. Given the new estimate of radiation-related fatal cancers, I would say that public exposure to radioactive gaseous or liquid releases and contamination of food must be kept below 0.6 mrem or 6 microsieverts per year. (Microsieverts are millionths of a Sievert.) This again calls for redesigning the industry and rethinking the permissible routine radionuclide releases. All of these changes involve major financial expenditures and full disclosure of problems to reactor purchasers.

One can only conclude that the assurances of safety for the nuclear industry were based on faulty atomic-bomb risk estimates and the difficulty inherent in proving liability for damage. Now, as the developed countries are faced with growing data on health damage, they appear to be dumping their unsafe technology on unsuspecting markets in the developing nations.

6. Nuclear Waste Isolation

As the hazard of ionizing radiation becomes more apparent, the engineering difficulty and expense of nuclear-waste isolation grows. Eventually, whatever is passively committed to land or sea in large quantities will make its way back into the food chain. The only controllable variable is the time line. Ethically, the question is: How far into the future do you care? Or, how many generations can demand our compassion?

It is my opinion that nuclear, and other toxic, waste needs to be secured and re-secured against release to the biosphere by each generation from now on into future time. The present state-of-the-art waste isolation does not warrant so-called passive placement.

Obviously, disposal of waste from the economically developed countries in Africa, South America, Asia or the Pacific-Island countries is unethical. The human costs of this colossal industrial mistake cannot be passed on to other nations and future generations.

7. Normal Operation or a Catastrophic Accident

Since a catastrophic nuclear accident can easily and dramatically outweigh a catastrophic accident at a coal, oil, or gas generator, we should consider the risks of both routine energy production, together with the risks of a catastrophic accident. The discussion of normal plant operation hinges on the estimate of detriment from routine worker and public exposure to radioactive releases. This has already been discussed in the previous section. On a short time scale, catastrophic accidents at a large hydro-electric dam may be more devastating than any nuclear scenario. However, common sense would dictate that persons assume complete responsibility for full damage due to an accident, regardless of its expression time.

A detailed risk analysis would need to be particularized to consider the location of an electrical power plant, the population density, geography of the area, farm land, and livestock at risk, as well as time into the future. Such an analysis would require more work than is possible or practical for this brief presentation. It must become part of each proposed project assessment.

Looking only at the historical record, under normal conditions of energy production and giving equal treatment to all energy back-up systems, the number of person-days lost due to death, illness, or accident, for each megawatt of electrical energy produced, is summarized in Table IV.[43]

With various degrees of incident, accident, or unplanned releases, the number of days lost associated with a nuclear source will increase. The newer (higher) risk factors have not yet been incorporated into this analysis.

Estimating the impact of a nuclear accident on public health is much more difficult. At Chernobyl, less than 5% of the radioactivity in the reactor was released. Estimates of cancer deaths in Europe due to that accident range from 100,000 to 600,000 over the next 50 years.

In an authoritative article published in *Science*, L. R. Anspaugh, of the Lawrence Livermore Nuclear Laboratory in California, estimated the number of cancer deaths for persons in the northern hemisphere, as a result of the Chernobyl accident.[44] His estimates are given in Table V. This more conservative estimate (derived by persons working for the nuclear military and commercial industries) just barely makes the Chernobyl reactor a "good neighbor" to the people of the Northern Hemisphere. The reactor caused less than one death per million persons per year. It is not a "good neighbor" by these standards for the whole of non-USSR Europe, however, because the non-fatal cancers would more than double these numbers.

The most worrisome part of a nuclear accident, however, is its ability to irreversibly damage the gene pool of humans and animals, producing a broad variety of illnesses and abnormalities. There is also grave danger from radiation-related depression of the immune system, which makes an individual susceptible to infectious diseases such as AIDS, TB, and meningitis. These problems are not even calculated in the estimates of "person days lost" for the normal-operation analysis. They may become the central and most important concern for Africa. The number of cancer deaths due to Chernobyl are only a small part of a sad story which involves poisoning of reindeer and caribou above the arctic circle, fish in the Nordic Peninsula, lambs in Great Britain, and grapes in Italy and Corsica.

The stakes in energy choices are very high, and the economics precarious. The ethical underpinnings of the nuclear industry are unacceptable. It is my sincere opinion that the current technology is fatally flawed, unacceptable in developed, and developing, countries alike. The major nuclear industry misjudgments of "acceptable" levels of radiation exposure will necessitate profound and expensive redesign of the technology or its abandonment.

TABLE I: FATAL CANCERS PER 10,000 PERSON-REM (OR 100-PERSON SIEVERT) PER YEAR

	ESTIMATE	RANGE
UNSCEAR (1977, 1986)	1	0.75-1.75
ICRP (1977)	1.25	----
BEIR I (1972)	1.17	2-5
BEIR III (1980)	0.8	----
BEIR V (1990)	7.6 males 8.1 females	
Hanford workers (1977)	26	2-49
John Gofman (1988)		
Atomic Bomb (1950-82)	41	33-44
Atomic Bomb (1950-85)	35	N.A.
Rosalie Bertell (1986)	10	5-16
Rudy Nusbaum (1989)		
Atomic Bomb (1950-82)	26	N.A.
Atomic Bomb (1950-85)	24	N.A.
D. L. Preston (1989)		
Atomic Bomb (1950-85)	17	N.A.
Linear Hypothesis (no threshold)	11	N.A.

TABLE II: COMPARISON WITH OTHER INDUSTRIES

	OCCUPATION	CAUSE OF FATALITY	DEATHS PER 10,000 WORKERS PER YEAR
(1)	Shoe Industry (Press & Finish)	Nasal Cancer	1.3
(2)	Printing	Lung Cancer	2.0
(3)	Workers with cutting oils	Cancer of scrotum	0.6-4.0
(4)	Wood machinists	Nasal cancer	7.0
(5)	Uranium mining	Lung cancer	15
(6)	Coal carbonizers	Bronchial cancer	28
(7)	Asbestos workers		
	Male (smokers)	Lung cancer	23
	Female (smokers)	Lung cancer	41
(8)	Rubber mill workers	Bladder cancer	65
(9)	Nuclear workers (at 20% permissible dose level)	All cancers (1990 estimates)	17-35
(10)	Nuclear workers (at 50% permissible dose level)	All cancers (1990 estimates)	42-88

Source: E. E. Pochin, "Occupational and Other Fatality Rates," *Community Health* 6, No. 2 (1974): 9. Nuclear-worker estimates were added, assuming the most recent atomic bomb estimates (1950-1985).

Table III: Industry-Reported Exposure Levels (USA)

Year	Reactor type	Average Exposure (rad or cGy)	% of Maximum Permissible Levels
1973	BWR	0.85	17
	PWR	1.00	20
1974	BWR	0.81	16
	PWR	0.68	14
1975	BWR	0.86	17
	PWR	0.76	15
1976	BWR	0.71	14
	PWR	0.79	16
1977	BWR	0.89	18
	PWR	0.65	13
1978	BWR	0.74	15
	PWR	0.65	13
1979	BWR	0.73	15
	PWR	0.55	11

TABLE IV: NUMBER OF PERSON-DAYS LOST PER YEAR

Coal fired	
Old technology:	20 to 86
New technology:	10 to 26
Oil fired	
Old technology:	6 to 44
New technology:	2 to 20
Natural Gas:	2 to 8
Nuclear:	10 to 236 (pre-1987 estimate)
Solar Thermal:	10 to 24
Solar Photovoltaic:	6 to 20
Solar Space Heating:	18 to 30
Wind:	16 to 17
Methanol:	10
Ocean Thermal:	4 to 6

Table V: Radiation-Induced Cancer Deaths due to the Chernobyl Accident (Lifetime)

Region	Rate Per million persons per year	Total number of deaths
USSR	0.3	6,500
Europe (non-USSR)	2.3	10,000
Asia (non-USSR)	0.004	500
USA and Canada	--.--	20
Northern Hemisphere	0.1	17,400

NOTES

1 W. C. Roesch (ed.), *U.S.-Japan Joint Reassessment of Atomic Bomb Radiation Dosimetry in Hiroshima and Nagasaki, Hiroshima*, U.S. NAS, RERF, 1987.

2 U.S. NAS, *Report of the Committee on the Biological Effects of Ionizing Radiation (BEIR V)*, Washington, D.C., U.S. National Academy Press, January 1990; hereafter cited as: BEIR V Report, 1990.

3 The ICRP draft version of its new recommendations was released in February 1990. The draft will be discussed at the main Commission meeting in June 1990. The Task Force drafting the final recommendations will meet in September 1990 and, if necessary, there will be another main Commission meeting in November 1990. The final text is expected in the first quarter of 1991.

4 Outlines of many alternative energy options are available from: Rocky Mountain Institute, 1739 Snowmass Creek Rd., Snowmass, Colorado 81654-9199 (Telephone: 303-927-3128).

5 A. Lovins, *Soft Energy Paths: Toward a Durable Peace*, Penguin, Harmondsworth, 1977.

6 See note 3.

7 D. L. Preston and D. A. Pierce, "The Effect of Changes in Dosimetry on Cancer Mortality Risk Estimates in the Atomic Bomb Survivors," Radiation Effects Research Foundation Technical Report TR 9-87, Hiroshima, Japan, RERF, 1987; hereafter cited as: Preston and Pierce, 1987. It should be noted

that 70% of the atomic-bomb-survivor population in the low-dose fraction are still alive. The total number of cancer deaths over a lifetime will increase.

8 BEIR V Report, 1990.

9 BEIR V Report, 1990.

10 Preston and Pierce, 1987.

11 R. H. Nussbaum, "Radiation Risks at Low Doses: Consistency Among Studies Previously Stamped As Incompatible," in *Proceedings of the International Conference on the Effects of Low Level Radiation on Man*, Un. of Munster (Fed. Rep. Germany) Feb. 26-27, 1988, Berlin, Springer Publ. Comp., 1989; hereafter cited as: Nussbaum, 1989.

12 R. Bertell, "Estimates of Uranium and Nuclear Radiation Casualties Attributable to Activities Since 1945," *Medicine and War* 4, (1988): 27-36; hereafter cited as: Bertell, 1988.

13 "The Danger of Radiation Has Been Underestimated," *Medical Post* 24 (19 January 1988).

14 K. Mather, Department of Genetics, University of Birmingham, U.K., Conference on the Biological Hazards of Atomic Energy, October 1950, in *Proceedings*, A. Haddow (ed.), Oxford Clarendon Press, 1952.

15 H. Muller, "Radiation and Heredity," *American Journal of Public Health* 54, No. 1 (1964): 42-50; hereafter cited as Muller, 1964.

16 Muller, 1964.

17 R. Bertell, "Radiation Exposure and Human Species Survival," *Environment Health Review* 25, No. 2 (1981): 43-52.

18 *Effect of Radiation on Human Heredity*, Report of a Study Group convened by the World Health Organization (WHO), 1957; hereafter cited as: WHO, 1957.

19 WHO, 1957.

20 BEIR V Report, 1990, p. 4.

21 Recommendations of the ICRP and lists of members have been regularly published by Pergamon Press (New York, London, Paris, Los Angeles) since 1959. Rules governing the organization appear first in 1959, ICRP #2. This report also points out record-keeping deficits.

22 U.S. HEW, *Statistics Needed for Determining the Effects of Environment on Health*, Report of the Technical Consultant Panel to the United States National Committee on Vital and Health Statistics, Washington, D.C., US, Department of Health, Education and Welfare, Series 4, No. 20. DHEW Publ. No. (HRA) 77-1457 (1977).

23 E. Rowman, *et al.*, "Childhood Leukemia in the West Berkshire and Basingstoke and North Hampshire District Health Authorities in Relation to Nuclear Establishments in the Vicinity," *British Medical Journal* 294 (1987): 597-602.

24 G. W. Kneale, *et al.*, "Hanford Radiation Study III: A Cohort Study of the Cancer Risks from Radiation to Workers at Hanford (1944-77 deaths) by the

Method of Regression Models in Life Tables," *British Journal of Industrial Medicine* 38 (1981): 156-166; hereafter cited as: Kneale, *et al.*, 1981.

25 A. Brodsky (ed.), *CRC Handbook of Radiation Management and Protection*, International Standard Book Number, 0-8493-3750-X, Section A, 0-8493-3755-0, Vol. II: Biological and Mathematical Information, 0-8493-3757-7, Boca Raton, Florida, CRC Press, 1982, p. 668.

26 No. 85-1095 in the Supreme Court of the United States of America, October 1985, Norman O. Aamodt and Marjorie M. Aamodt vs. U.S. Nuclear Regulatory Commission, On Writ of Certiorari to U.S. Court of Appeals for the Third Circuit (Appeal nos. 85-3301, 3302, 3310, 3315). There were also a large number of court claims settled out of court by Metropolitan Edison Co.

27 WHO, 1957.

28 D. Black, *et al.*, "Investigation of the Possible Increased Incidence of Cancer in West Cumbria," Report of the Independent Advisory Group, Chairperson, Sir Douglas Black, London, U.K., 1984.

29 R. Bertell, "Environmental Influence on Survival of Low Birth Weight Infants in Wisconsin 1963-75," *International Perspectives in Public Health* 2 (1984).

30 R. Bertell, "Radioactive Residue in Residential Areas," (1986). Never published, by now incorporated into a larger paper to be submitted for publication.

31 *United Nations Scientific Committee on the Effects of Atomic Radiation* (UNSCEAR) Report, New York, U.N., 1977.

32 Bertell, 1988.

33 J. Sevec, *et al.*, "Lung Cancer in Uranium Miners and Long Term Exposure to Radon Daughter Products," *Health Physics Journal* 30 (1976): 433-437.

34 U.S. NAS, *BEIR III 1980*, Washington, D.C., U.S. National Academy of Sciences, pp. 318-320.

35 *International Atomic Energy Agency Safety Guides*, Series No. 89, 1988, p. 8.

36 See note 3.

37 R. Bertell, *Handbook for Estimating Health Effects from Exposure to Ionizing Radiation*, 2nd Revised Edition, Toronto, International Institute of Concern for Public Health, 1986; hereafter cited as: Bertell, 1986.

38 Kneale, *et al.*, 1981.

39 Bertell, 1986.

40 Nussbaum, 1989.

41 Entire study available in German from B. Franke, Institute fur Energie und Umweltforschung, Heidelberg, F.R.G. See Bernd Franke, *et al.*, "Risikoorientierte Analyse zum SNR 300 bei Kalkar," Heidelberg, Februar 1982, IFEU, Umweltforschung, Heidelberg, e.V., 6900 Heidelberg 1, F.R.G.

42 See note 3.

[43] L. R. Anspaugh, *et al.*, "The Global Impact of the Chernobyl Reactor Accident," *Science* 242 (December 16, 1988): 1513-1519; hereafter cited as: Anspaugh, 1988. For further information, see: R. Bertell, "The Nuclear Worker and Ionizing Radiation," *American Industrial Hygiene Assoc. Journal* 40 (1979): 395-401. R. Bertell, N. Jacobson, and M. Stogre, "Environmental Influence on Survival of Low Birthweight Infants in Wisconsin 1963-75," *Intr'l Perspectives in Public Health* 1 (1984). Karl Z. Morgan, "Hazards of Low-Level Radiation," *Yearbook of Science and the Future*, Supplement of the Encyclopedia Britannica, 1980. Robin Russell Jones and Richard Southwood (eds.), *Radiation and Health*, New York, John Wiley and Sons, 1987. On ethical aspects, also see: R. Bertell, "Humanity at Risk: Current Challenges to the Christian Scientist," *Canadian Catholic Review* (1984): 441-447. K. S. Shrader-Frechette, *Nuclear Power and Public Policy, The Social and Ethical Problems of Fission Technology*, Boston, Reidel, 1983.

[44] Anspaugh, 1988.

Nuclear Wastes and Ethics

Kristin Shrader-Frechette
University of South Florida
Tampa, Florida, USA

Egyptians have been unable to protect the tombs of the Pharoahs for less than *four* thousand years, and some of them were looted within centuries. Italians have been unable to protect Renaissance art treasures for less than *one* thousand years.[1] Yet we in this generation have to be able to protect nuclear wastes, for *hundreds* of thousands to millions of years, for a period longer than all recorded history, because of their long-lived radionuclides (like plutonium 239, carbon 14, and iodine 129)[2]

There are no plausible inductive arguments based on data for radwaste storage for *one* century, much less data for the minimum storage period of *thousands* of centuries. What data we do have is not comforting. Regulatory problems and safety questions have repeatedly forced the Energy Department to delay planning and opening facilities for nuclear waste. And, in the past, hundreds of people have died in radwaste accidents. The worst occurred when thousands of acres - 22 square miles - in Soviet Kasli were rendered uninhabitable by high-level radwaste that went critical three decades ago.[3] And, at the premier US high-level facility, in Hanford, Washington, over 500,000 gallons of high-level radwaste have leaked into the soil, then into the Columbia River, and then into the Pacific Ocean.[4] Despite such radwaste accidents, nuclear proponents continue blindly to assert that "none of the cases reported appears to have led to a radiological hazard."[5]

At the nuclear-waste facility containing more plutonium than any other commercial site in the world, Maxey Flats, experts were wrong by six orders of magnitude, when they predicted how fast the stored plutonium would migrate. They said that it would take 24,000 years for the plutonium to travel one-half inch onsite. It went two miles offsite in ten years. Current plans for future US storage of high-level radioactive waste require the steel canisters to resist corrosion for as little as 300 years. Nevertheless, the US Department of Energy admits that the waste will remain dangerous for longer than 10,000 years. Government experts agree that, at best, they can merely limit the radioactivity that reaches the environment, and that "there is no doubt that the repository will leak over the course of the next 10,000 years."[6]

The US government has extrapolated, on the basis of past leaks at its nuclear waste facilities, and has said that future leaks should occur at a rate of two to three per year. Using (US) government-estimated exposure levels (580 person rem) at each radwaste site, each existing facility could cause approximately 12 cancers and 116 genetic deaths per century.[7]

These numbers appear relatively small until one realizes that, for the period of storage required, the cancer deaths alone would be in the tens of thousands per storage site. Despite such government data, nuclear proponents continue to assert that "the radiological impact of waste repositories will be small."[8] Admittedly there are new technologies that might reduce these government figures. Any new technology, however, is unavoidably dependent upon fragile and short-lived human institutions and human capabilities. It wasn't faulty technology, after all, that caused Three Mile Island or Chernobyl. It was human error. Likewise it could well be human error that is the insoluble problem with managing radwaste.

Perhaps because they are worried about human error and about scientists' claims to store waste safely in perpetuity, citizens, environmental groups, and states have been in turmoil over the issue of radioactive waste storage. Virtually no one wants it in her backyard. Numerous groups have charged in lawsuits that radiation exposure standards for proposed sites are too lax and violate the Safe Drinking Water Act. Over the last two years, the US Congress has been besieged with more than 30 bills proposing to delay, abandon, or change the repository program established under the 1983 US Nuclear Waste Policy Act.[9]

In the absence of certainty about our ability to store the wastes, we need ethical criteria for accepting or rejecting the technological risks they impose. To develop some ethical criteria, we look first and briefly at the magnitude of the radwaste problem. Second, we survey some of the technical, epistemological, and ethical problems raised by high-level radwaste storage. That accomplished, we outline four philosophical dilemmas posed by radwaste. The final section sketches some solutions to the difficulties that we have surveyed.

1. A Revisionist History of Commercial Nuclear Fission

Standard accounts of nuclear history are often misleading regarding the source and the magnitude of our difficulties with radioactive waste. Some persons have claimed that we face the problem of nuclear waste both because of military activities and important hospital uses of nuclear medicine and because a number of utilities were eager to provide inexpensive electricity for a great number of people. Both these myths are untrue.

High-level radwaste is, for the most part, spent fuel rods from fission reactors and residues from fuel reprocessing.[10] Less than one percent of high-level radwaste is from medical activities.[11] Moreover, at least in the US, approximately half of the high-level radwaste now needing storage is from commercial nuclear fission, not military activities.[12] Even more important, the commercial half of the waste, primarily spent fuel, is expected to rise dramatically by 1995, to 11 times the metric tons now needing to be stored, while the military waste will increase very slowly and remain close to current levels. For example, US spent fuel, currently about 7,000 metric tons, will rise to more than 70,000 metric tons by 1995. The upshot is that, by 1995, most high-level and low-level radwaste will be from commercial reactors, not military activities, and certainly not medical processes.[13]

Nor do we have the problem of nuclear waste because industry was eager to generate electricity, and fission was an economical means of doing so. At the beginning of the atomic era, industry was reluctant, both on economic and on safety grounds, to use fission to generate electricity. Worried about safety, every major US corporation with nuclear interests refused to generate nuclear electricity unless some indemnity legislation was passed to protect them, in the event of a major accident.[14]

The top industry lobbyist for nuclear fission, the President of the Atomic Industrial Forum, has confirmed what numerous government committee reports from the forties and fifties revealed. Commercial nuclear fission began, and was pursued, only because government hoped to justify continuing military expenditures in nuclear-related areas and to obtain weapons-grade plutonium.[15] Moreover, at least in the US, nuclear fission began only because government provided more than $100 billion in subsidies (for research, development, waste storage, and insurance) to the nuclear industry. It also gave the utilities a liability limit (in the Price-Anderson Act) that protects licensees from 99 percent of all public losses and claims in the event of a catastrophic nuclear accident.[16]

Twenty years after commercial fission reactors began operating, in 1976, the *Wall Street Journal* proclaimed them an economic disaster. Nuclear electricity has proved so costly that year 2000 projections for commercial fission reactors are now approximately one eighth of what they were in the mid seventies. No new reactors have been ordered in the US since 1974.[17]

The few US nuclear manufacturers that are still in business have remained by selling reactors to other nations, often developing countries. Yet, many of the commercial reactors going to these nations may not be in their best interests, as later sections of this essay will suggest. Indeed, commercial nuclear fission may be the current version of infant formula. In the infant-formula controversy of the last two decades, US and multinational corporations made great profits by exporting infant formula to developing nations. They were able to do so only by coercive sales tactics and by misleading foreign consumers about the merits of their products.

Some diplomats also have charged that developing nations are seeking fission-generated electricity as a subterfuge for obtaining nuclear-weapons capability,[18] through the plutonium byproduct. India exploded its first nuclear bomb, for example, by using plutonium produced by a reactor exported by Canada. Whether or not this is the reason for the survival of nuclear fission, the few countries still developing atomic energy do not have uncontroversial grounds for subscribing to the myth of the safe, economical atom. Fission is not obviously a cheap, sensible, or ethical way to boil water. As subsequent remarks shall illustrate, there are better ways to run a turbine.

2. Technical Problems with High-Level Radwaste Storage

With this revisionist nuclear history behind us, let's look at some of the main technical problems posed by radwaste, especially high-level radwaste. Each year, each 1000-MWe reactor discharges about 25.4 metric tons of high-level waste, spent fuel.[19] For 300 commercial reactors, world-wide, the annual high-level radwaste would be 7620 metric tons per year. Compare this to the fact that 10 micrograms of plutonium is almost certain to induce cancer, and that several grams of plutonium, dispersed in a ventilation system, are enough to cause thousands of deaths.[20] Moreover, each of the 7620 metric tons of high-level waste produced annually has the potential to cause hundreds of millions of cancers for at least the first 300 years of storage, and then tens of millions of cancers for the next million years.[21]

These cancers could be prevented, of course, with perfect isolation of the wastes for a million years. That is why most plans for high-level radwaste storage call for defense in depth, for sealing the waste in a ceramic material, and for burying it deep underground. Despite these plans, the US Environmental Protection Agency (EPA) has warned that we cannot count on institutional safeguards for the waste beyond 100 years.[22]

Even if there were institutional safeguards, however, the geological and hydrological problems would remain. For example, many people have proposed that high-level radwaste be stored in salt deposits. Yet salt deposits are often located near valuable mineral deposits (e.g., potash in New Mexico and natural gas in Louisiana and New York) where drilling is likely to occur. Moreover, high-level radwaste is a heat source, and when salt is heated, water is attracted to it. The resulting brine could eventually break down either canisters or ceramics of high-level radwaste. The US Geological Survey (USGS) has noted that increased temperatures and the presence of brine would undermine the structural integrity of the salt repository and would increase the creep rate of dry salt. This leaching and creeping, in turn, could cause migration of the waste, since most salt deposits lie above limestone.[23]

In other words, as the famous US Interagency Review Group on nuclear-waste management reported: the scientific feasibility of dry storage in geologic repositories, deep in salt beds, or hard rock, "remains to be established."[24] As a result, granite storage sites in Sweden have been vetoed as unsafe, Kansas salt beds have been rejected because they are riddled with holes, and the first model US repository for high-level radwaste will not be ready before 2008 or later.[25]

In the absence of proof that we can successfully store radwaste, the technical problems associated with it are forcing us to take a great gamble. This is a gamble that our descendants will not breach the repositories through war, terrorism, or drilling for minerals; a gamble that water and heat will not combine to create nuclear reactors in underground waste, as already happended in the USSR; and a gamble that ice sheets and geological folding will not uncover the wastes.

Scientists have not been worried by these gambles because, for the first 35 years of commercial nuclear fission, they were saying that safely isolating the wastes would be easy, once they set their minds to the task. They have been like contractors who built houses without toilets,[26] and then alleged that constructing the toilets would be easy. Since no country has a permanent high-level disposal facility,[27] perhaps the task is not so easy as has been alleged.

3. Epistemological Problems with Radwaste Storage

Even more staggering than the technical obstacles are the philosophical issues raised by radwaste. Most of the epistemological difficulties arise from the fact that secure storage cannot be guaranteed. Hence, regardless of the technology used, anyone who favors a particular method of radwaste management must use some form of the fallacy of the appeal to ignorance. Namely, "I know of no way in which containment could be breached; therefore containment will probably not be breached." Without falling into logical fallacy, one cannot draw a positive conclusion about radwaste safety when the premises upon which it is based have so little empirical foundation.

Of course, scientists are often forced to draw conclusions without empirical tests, and much progress in science consists of theorizing on the basis of slim empirical foundations. Because this highly theoretical approach is legitimate in science, however, it is not clear that it is applicable to a technology capable of putting millions of lives at risk. We know that, in the past, we were wrong to use an appeal to ignorance when we dumped radwaste into the sea, when we treated mastitis with radiation, when we used x-rays to determine shoe fit, and when we subjected US soldiers to nuclear-test fallout during peacetime. We may likewise be wrong to use an argument from ignorance to conclude that we can isolate radwastes in perpetuity.

Another epistemological problem is how we can completely isolate the radwaste from the biosphere and yet monitor it to insure that containment has not been breached. Complete isolation appears to preclude adequate monitoring, and adequate monitoring appears to preclude isolation.

A third epistemological problem is how to guarantee, scientifically and technologically, the so-called "neutrality criterion." This criterion specifies that the levels of risk to which future generations will be subjected, because of the waste, must be no greater than those of present persons.[28] Because of the absence of good inductive evidence, any suggestion that this criterion can be met amounts to another argument from ignorance. Nuclear proponents almost always claim that risks to future persons from radwaste will be be greater than those to present persons. Yet, at the same time, they espouse a "dilution and dispersal" concept of waste management.[29] Logically, they cannot argue for continued dilution and dispersal, and yet maintain that risks in the future are not continually increasing. Other assumptions - that the future will be geologically, politically, and institutionally stable enough to insure that current plans for radwaste management are not disrupted - likewise amount to little more than arguments from ignorance.[30] Perhaps the greatest argument from ignorance is nuclear proponents' continued use of the term, "radwaste disposal."[31] If some leakage is unavoidable, and if radiation exposures have no threshold for harm and are cumulative, then there is no "disposal"- only "management" or "storage."

4. Ethical Problems with Radwaste Storage

Most of the major ethical problems associated with radioactive waste focus on the issue of equity. Roger Kasperson, in his famous study of radwaste management, places them in three main groups: locus problems, legacy problems, and labor-laity problems. The locus issues have to do with where and how to site radwaste facilities. The labor-laity problems focus on whether to maximize either the safety of the public or that of radwaste workers, since both cannot be acomplished at once. The legacy problems have to to with exporting radwaste risks and costs to future generations.[32]

The key question raised by legacy concerns is whether one can justify intergenerational inequity by mortgaging the future, by imposing our debts of radioactive waste on subsequent generations. If we saddle our descendants with our medical and financial debts of waste, then taxpayers in later centuries could be forced to pay an annual tab for radwaste storage between $3.8 and $1.9 million per reactor per year.[33] This expenditure is obviously questionable because future generations ought not be saddled with other persons' debts. Also, there is very little public funding, especially by taxpayers, for decentralized energy technologies, like solar, which are much less likely to burden the future.[34]

A second legacy question is what sort of criteria might justify environmentally irreversible damage to the environment, like that caused by deep-well storage of high-level radwaste. Radwaste management schemes which are irreversible theoretically impose fewer management burdens on later generations, but they also preempt future choices about how to deal with the waste. On the other hand, schemes which are reversible allow for greater choices for future generations, but they also impose greater management burdens. If we can't do both, is it ethically desirable to maximize future freedom or to minimize future burdens?

Perhaps the most important legacy question concerns the contribution of radwaste production and storage to the "plutonium economy" which is necessary for building nuclear weapons. Would our continuing to generate and store radwaste maximize or minimize future freedom and security? Would it add to the probability of a nuclear holocaust by providing materials necessary for arms?[35]

Still other legacy questions have to do with use of social discount rates. Any alleged economies or safety associated with high-level radwaste storage are in large part questionable because of their dependence on a particular discount rate. Using a discount rate amounts to discounting future costs (of radwaste storage, like radiation-related deaths or injuries) at some rate of *x* percent per year. Thus at a discount rate of 10 percent, effects on people's welfare twenty years from now count only for one-tenth of what effects on people's welfare count for now. With a discount rate of 5 percent, effects next year count for 1,000 times more than effects 200 years from now. Or, more graphically, with a discount rate of 5 percent, a billion deaths in 400 years counts the same as one death next year. A number of famous moral philosophers, like Parfit, have argued that use of a discount rate is unethical, because the moral importance of future events, like the death of a person, does not decline at some *x* percent per year.[36] Yet radwaste storage is typically said to be economical only because costs borne by future persons are discounted. Without discounting, it would be virtually impossible to justify the dangers and costs of storing radwaste for thousands of centuries.

Imposing nuclear waste on future generations might also be questionable from a practical point of view. Uranium will not be available much beyond the year 2000, even to supply existing fission reactors. Hence, after having generated tons of highly toxic waste, nuclear energy (without the breeder and without fusion) will not have provided a long-term technological fix for our energy problems. From a practical point of view, it is not clear that the *temporary* benefits of nuclear fission are worth the *permanent* costs of radioactive waste.[37]

If there were no alternatives to our using nuclear power, then obviously we could justify imposing inequitable, and perhaps impractical, radwaste risks on the future. But to the degree that there are alternatives to commercial fission generation of electricity, then to that extent is the imposition unethical. Later in the discussion, we shall address some of these energy alternatives.

In addition to the legacy issues, there are a number of locus or siting problems associated with managing radioactive waste. Consider, following Locke, Blackstone, Nozick, and others, that it may be impossible to justify acquisition of property rights in land, especially original property rights. If one has rights only to what is produced by one's labor, and if one's labor has not produced land, then it is questionable whether one might be said to have full property rights to it.[38] But if so, then how does one justify siting any land for radwaste storage? This is a use which amounts to exercising the most extreme form of property rights, since it preempts both present and future choices about land use at that site.

Another important siting or locus issue is geographical equity, especially since it is a foregone conclusion that radwaste depositories will be located in rural areas, away from major population centers.[39] One of the questions raised by considerations of geographical equity is whether it is fair to impose a risk on a person just because she lives in a rural community rather than a large city. Is it just to impose a higher risk of death/injury on a person just because she lives in a waste-storing, rather than a waste-producing, community?

Likewise, it is ethical for one geographical subset of persons to receive the benefits of nuclear-generated electricity, while a much larger set of persons bears the costs? This particular question of geographical equity arises because of the peculiarities associated with taxpayer funding for radwaste storage. Despite the fact that utilites contribute ($1 for every 1,000 kw hours of electricity that they generate) toward a radwaste management fund,[40] these contributions cover only small fraction of storage costs. The bulk of storage expenditures is not included in calculations of the cost of nuclear electricity. Cost comparisons of alternative energy technologies typically ignore externalities like the $100 billion in nuclear subsidies already spent by US taxpayers, part of which is for nuclear waste storage.[41]

In the past decade, for example, government subsidies to the nuclear industry, in the form of write-off for capital invested in plants not completed, were $4 billion in the US alone.[42] It is arguably inequitable to ignore subsidies paid by all the taxpayers to an industry that primarily benefits only a small set of taxpayers.

Once taxpayer subsidies for costs like waste storage and decommissioning are included in the calculations, nuclear power can be shown to be more expensive than every other energy alternative.[43] The only way to make it viable is to remove it entirely from the discipline of the market, and to increase taxpayer subsidies.[44] Already this removal from the market has resulted in members of future generations, and taxpayers generally, picking up the tab for billions of dollars of expenses that arguably should be borne by the nuclear industry and its beneficiaries. Hence radwaste funding raises a whole host of equity questions that affect both locus and legacy.

5. Philosophical Dilemmas

Rather than discuss each of these ethical problems, a task beyond the scope of a short chapter, let us look at four classical philosophical dilemmas posed by radwaste storage. These four dilemmas should be sufficient to provide some insight into possible solutions to some of the locus, legacy, and laity problems raised by radwaste. I call them the consent dilemma, the federalism dilemma, the threshold dilemma, and the contributors dilemma.

5.1 *The Consent Dilemma*

The Consent Dilemma is that siting radwaste facilities and employing waste management workers requires the consent of those put at risk, yet those most able to give free, informed consent are usually unwilling to do so, and those least able to validly consent are often willing to give alleged consent. Hence it is difficult, if not impossible, to achieve free, informed consent to risks like those posed by radioactive waste.

To see how the consent dilemma arises, consider a typical case. When West Valley, New York was proposed as a storage site for low-level radioactive waste, townspeople were eager to obtain the economic benefits they believed the facility would bring. City leaders predicted a "boomtown," and this was the basis of community acceptance of the site. Although the boomtown never occured, the facility paid 20 percent of the town and county taxes. From a superficial perspective, it might look as if West Valley citizens gave free, informed consent to the waste site. But consider what sorts of communities would be most likely to consent to a radwaste facility nearby. Probably not those whose residents had high income, job security, and high levels of education. Communities full of people with these characteristics would know enough to be wary of the risk, and they probably would already have excellent jobs. Hence they would not need to take any risks in order to better themselves economically.[45]

Or consider the case of workers at another radwaste facility. A British study of 35,000 living workers from the Hanford high-level radwaste facility showed increased chromosomal damage in workers exposed to less than one-half of the annual allowable exposure of 5 rems of radiation.[46] Workers at the facility were justified in having these risks imposed on them, according to classical economic theory, because of an alleged compensating wage differential. According to the theory behind the alleged differential, the riskier the occupation, the higher the wage required to compensate the worker for bearing the risk, all things being equal.[47]

According to classical ethical theory, imposition of these higher workplace risks is legitimate apparently only after the worker consents, with knowledge of the risks involved, to perform the work for the agreed-upon wage. But the dilemma arises once one considers who is most likely to give legitimate informed consent. It is a person who is well educated and adequately informed about the risk, especially its long-term and probabilistic effects. It is a person who is not forced, under dire financial constraints, to take a job which he knows is likely to harm him.

Yet, sociological data reveal that, as education and income rise, persons are less willing to take risky jobs, and those who do so are primarily those who are poorly educated or financially strapped. Sociological data also shows that the alleged compensating wage differential does not operate for poor, unskilled, minority, or non-unionized workers. Yet these are precisely the persons most likely to work at risky jobs like storing radwaste.[48] This means that the very set of persons *least* able to give free, informed consent to workplace radwaste risks are precisely those who most *often* work in risky jobs and are alleged to have given consent.

If this observation about worker radwaste risk and community acceptance of a radwaste site is accurate, then medical experimentation may have something to teach us about risk assessment. We know that the promise of early release for a prisoner who consents to risky medical experimentation provides a highly coercive context which could jeopardize her legitimate consent. So also high wages for a desperate worker who consents to take a risky job provides a highly coercive context which could jeopardize her legitimate consent. This means that we must either admit that our classical ethical theory of free, informed consent is wrong, or we must question whether those closely affected by radwaste risk genuinely consent to it.

5.2 The Federalism Dilemma

Consider now a second classic dilemma. Liberty and grass-roots self determination require local control of whether a radwaste facility is sited in a particular area. Yet, equality of consideration for people in all locales, and minimizing overall risk, require federal control. Do we say that the local community can veto a radwaste site, even though that site may be the best in the country and may provide for the most equal protection of all people? Or do we say that the national government can impose a radwaste site on a local community, even though the imposition is at odds with their free and self- determined choice?

This particular dilemma is all the more pressing both because few communities seem to want radwaste, and because the issue of national supremacy versus local autonomy has created lively debate for several centuries. On the one hand, national supremacy is likely to protect the environment, to avoid the tragedy of the commons, to gain national economies of scale, and to avoid regional disparities in representing all sides in the radwaste debate. National supremacy also is likely to provide for compensation of the victims of one region for spillovers from another locale and to facilitate the politics of sacrifice by imposing equal burdens on all.[49]

On the other hand, local autonomy is likely to promote diversity, to offer a more flexible vehicle for experimenting with radwaste regulations, and to enhance citizens' autonomy and liberty. It also is likely to encourage fraternity through participation in decisionmaking, to avoid inequitable federal policies, and to avoid violations of rights.[50]

How do we resolve this dilemma? Noted law professor, R. B. Stewart, claims that it is impossible to maximize both liberty and environmental integrity, or both liberty and equality. The dilemma cannot be resolved in the sense of maximizing all three values.[51]

5.3 *The Threshold Dilemma*

A third problem, the threshold dilemma, is based on the fact that society must declare some threshold below which risk is declared to be negligible or minimal, so far as its acceptability is concerned. Typically this threshold level is set at what would cause less than a 10^{-6} increase in one's average annual probability of fatality.[52]

The reasoning behind setting such a level is that a zero-risk society is impossible, and some standard needs to be set, especially in order to determine pollution-control expenditures. Choosing the 10^{-6} standard also appears reasonable, both because society must attempt to reduce larger risks first, and because 10^{-6} is the natural-hazards death rate.[53]

The dilemma arises because no threshold standard is able to provide *equal* protection from harms like radwaste to all citizens. Any such standard guarantees merely that an *average* annual probability of fatality is associated with some hazard. Because this 10^{-6} threshold seems acceptable on the average, however, does not mean that it is acceptable to each individual. Blacks, for example, face higher risks from air pollution than do whites, even though they share the same "average" exposure.[54] Likewise those around radwaste facilities will be exposed to radiation levels for average persons, not for them as individuals with unique needs.

Most civil rights, however, are not accorded on the basis of the *average* needs of persons, but on the basis of *individual* characteristics. For example, we do not accord constitutionally guaranteed civil rights to public education, on the basis of average characteristics of students. If we did, then retarded children or gifted children would have rights only to education for children at the average level. Instead, we say that according "equal" civil rights to education means according "comparable education," given one's aptitudes and needs. That is why the state can provide special schools for both the retarded and the gifted.

This example from the field of education raises an interesting question for radwaste risks and indeed for all risks requiring a threshold to be set: If civil rights to education are accorded on the basis of individual, not average, characteristics, then why are civil rights to equal protection from risks not accorded on the basis of individual, rather than average, characteristics? Why is a 10^{-6} average threshold accepted for everyone, without compensation, when adopting it poses risks higher than 10^{-6} for the elderly, for children, for persons with previous exposures to radiation, for those with allergies, for persons who must lead sedentary lives, and for the poor?

In dealing with this dilemma, one is forced to make a value judgment, either to adopt a particular threshold standard, or a more stringent standard that would afford more equal protection for everyone. Which shall we choose: average protection or equal protection, efficiency in regulation or equality of protection?

5.4 *The Contributors Dilemma*

A fourth difficulty is what I call the Contributors Dilemma. It consists of the fact that citizens are subject to numerous small risks, e.g., to certain carcinogens, each of which is allegedly acceptable; yet, together such exposures are clearly unacceptable. In the case of radioactive waste, selected groups of citizens, such as those living near a storage facility or a radwaste transport route, are exposed to radiation. Each of these small exposures is alleged to be acceptable because it is below the threshold at which some statistically significant increase in harm occurs, yet together these exposures can cause serious damage.

Statistically speaking, 25 to 33 percent of us are going to die from cancers. The US Office of Technology Assessment says that 90 percent of these cancers are environmentally induced and hence theoretically preventable.[55] Many of the cancers are obviously caused by the aggregation of numerous exposures to carcinogens, like radiation, no one exposure of which is alone alleged to be harmful.

The Contributors Dilemma is especially problematic in the case of devising ethical regulations for small risks, like radiation. Risk assessors who condone sub-threshold risks, but who condemn the deaths caused by the aggregate of these sub-threshold risks, are something like the bandits who eat the tribesmen's lunches in the famous story of Jonathan Glover:

> Suppose a village contains 100 unarmed tribesmen eating their lunch. 100 hungry armed bandits descend on the village and each bandit at gun-point takes one tribesman's lunch and eats it. The bandits then go off, each one having done a discriminable amount of harm to a single tribesman. Next week, the bandits are tempted to do the same thing again, but are troubled by new-found doubts about the morality of such a raid. Their doubts are put to rest by one of their number [a government risk assessor]....They then raid the village, tie up the tribesmen, and look at their lunches. As expected, each bowl of food contains 100 baked beans....Instead of each bandit eating a single plateful as last week, each [of the 100 bandits] takes one bean from each [of the 100] plate[s]. They leave after eating all the beans, pleased to have done no harm, as each has done no more than sub-threshold harm to each person.[56]

The obvious question raised by this example is how a risk assessor can say both that sub-threshold radwaste exposures are harmless, as the data allegedly indicate, and yet that the additivity, or contribution, of these doses causes great harm. It appears that risk regulators need to amend their theory regarding synergistic or additive risks like cancer.

6. Two Solutions Required by Ethics: Consent and Compensation

Although there is inadequate space here to defend the sorts of solutions that would help resolve these four dilemmas, it is possible to outline briefly the arguments which, when developed, would provide such a defense. One solution to the consent dilemma and to the federalism dilemma is to apply the same strict standards of informed consent, already well known in doctor-patient or experimenter-patient relationships, to the radwaste siting issue. Indeed they could be applied to all matters of technological controvery where consent is relevant.

We all probably believe that it would be unethical for a doctor to impose a risky treatment on a patient without her free, informed consent. Yet most of us need to develop our moral sensibilities so as to see radwaste workers or radwaste siting in the same light. Until or unless a risk imposition receives the consent of those who are its potential victims, it cannot be justified. This means that those who wish to impose societal risks on others need to do whatever is necessary to compensate them to a degree adequate to obtain their free, informed consent. If no forms of compensation are adequate to obtain free, informed consent, then it is questionable whether the risk imposition can be justified.

Whenever the federalism and consent dilemmas have arisen in radwaste siting issues, these have often been resolved by means of increasing the compensation to the risk takers likely to be affected by the waste site. In some cases in which it was not possible to meet the compensatory demands of a community proposed as a radwaste site, the facility was not located. In the cases in which compensation was able to be negotiated, however, often this was sufficient to insure community consent. Typically industry-supplied compensation in radwaste cases includes tax breaks or funding community projects or schools. According to social scientists (studying waste siting), however, one form of compensation dominates all negotiation. The one factor almost always essential to achieving local consent is giving citizens/workers funding to control health and safety monitoring themselves at the facility. By forcing the waste managers to pay for outside monitors, citizens and workers are freed from relying on company monitoring. Moreover, once they have greater control of their health and safety, they appear ready to give informed consent to the risk.[57]

Compensation also provides one solution to the threshold dilemma and to the contributors dilemma, both of which raise equity issues. By providing full compensating benefits for all unavoidable radwaste risks, industry can offset the inequities generated by some of the locus and legacy problems.[58]

Admittedly compensation of future generations who bear the brunt of the legacy problems might be difficult. Obtaining free, informed consent and guaranteeing compensation in such cases requires that the consent of future persons be obtained by means of representatives acting in their best interests. It also requires that those who store radioactive waste actually set up a fund for compensation of future persons possibly harmed by the waste.

Requiring full compensation and consent in the case of future generations entails, in part, that we use an equal-opportunity criterion for intergenerational equity. To the degree that we cannot guarantee that persons a million years from now will have equal opportunity to protect themselves from our radwaste hazards, then to that same degree we cannot justify generating radioactive waste or imposing it upon others.[59]

The rationale for requiring full compensation and consent, even regarding future persons, is that any technology ought to "pay its own way." Considerations of fairness dictate that a technology ought not be subsidized by the deaths and cancers of persons who are its unwilling victims. If nuclear power cannot "pay its own way" regarding compensation and consent, then it has no right to claim that it ought to be accepted. To the degree that nuclear power does not pay its own way, especially in regard to waste storage, it will continue to reinforce all the old income distributions and inequities. It will reaffirm a situation in which the poor and the uneducated bear the social costs of contemporary technology.

Also, if someone can impose a bodily risk of harm on another without his consent, then there is no right to life. If someone can profit by imposing a threat of physical harm on another without compensating him, then there is no right to due process. These considerations suggest that, if we refrain from requiring genuine informed consent and from guaranteeing complete compensation for radwaste risks, then radwaste will put more at risk than our health and that of future generations. It will put at risk our most basic rights to justice and equity.

7. These Ethical Solutions Make Economic Sense

Providing equity, informed consent, and full compensation regarding the risks associated with managing radioactive waste does not mean, however, that we must forego common sense in order to sustain our ethical ideals. Foregoing use of nuclear power and the generation of additional radwastes might also be an economically sound decision, for a variety of reasons.

As was already mentioned earlier, virtually all commercial nuclear construction in developed nations has come to a halt, largely because fission generation of electricity is uneconomical and likely unsafe. France is the only developed country with an ongoing nuclear program, allegedly the most successful in the world. This means that, to address the question of the economics associated with commercial nuclear fission, one needs to examine primarily the French case and that of developing nations. The issue has largely been decided (against nuclear fission) for other (developed) countries.

Although the French nuclear program is widely viewed as successful, several considerations suggest that this description may not be accurate. For one thing, the French use a breeder reactor, not the fission technology employed in other energy programs around the world. Most nations have decided against the breeder because it generates inordinate amounts of radioactive waste and hence creates a "plutonium economy" that creates vast amounts of materials for nuclear weapons. It also poses a waste-storage problem several orders of magnitude greater than that created by fission reactors.

Moreover, the apparent reasons for the success of the French nuclear program - the fact that it is centralized and government owned - are artificial and not transferrable to other nations. The French utility is protected from market forces, protected from environmentalists' criticisms, and protected from public participation in decisionmaking. Even with these benefits, the French utility now has a debt of $30 billion, due to its commitment to nuclear power. In 1982 this debt had accumulated to $152 billion, although part of it was forgiven by the government. In light of the French deficit, developing nations, many of which are on the edge of insolvency, will be hard pressed to pay the enormous capital costs for nuclear reactors.[60] Since developed nations, except for France, have cut back or halted their fission programs, this means that the commercial viability of nuclear power is an issue only for developing nations.

Developing nations, in particular, will probably be hard pressed to pay the costs of decommissioning and waste storage associated with their approximately 30 nuclear reactors. Some experts claim that waste-storage costs will add an estimated 5-10 percent to the total cost of nuclear power.[61] US industry experts maintain that decommissioning a reactor could alone amount to 24 percent of original construction costs. This translates to more than $500 million for recently built facilities. If French nuclear industry experts are correct, then the cost would be even higher. They say that decommissioning currently would run at least 40 percent of initial construction costs of the reactor.[62]

Moreover, apart from paying for decommissioning the reactor and managing the waste, most nations (and especially developing ones) will likely have a difficult time paying for nuclear power at all. Since the middle seventies, nuclear-plant construction costs have doubled every four years. More than 25 percent of these costs are for financing.[63] Part of the reason for these increases is that, because of additional safety measures needed in the wake of accidents, the amount of concrete, piping, and cable used in the average plant has more than doubled, while labor requirements have more than tripled.[64] No reasonable utility owner in any country is going to build something, a nuclear plant, when she doesn't even know its full costs, especially decommissioning and waste storage.

A particular obstacle to nuclear plants in developing nations is the small size of electricity grids in most of them. If a single power plant provides more than 15 percent of the grid's capacity, the whole system will "crash" if that plant is shut down. Yet only four or five developing countries have grids large enough for a conventional 1,000 MWe reactor. This problem could be addressed by plants smaller than the conventional 1,000 MWe ones, but the per-kilowatt construction cost for a nuclear plant of 200 MWe is more than twice that for a 1,000 MWe one, and even the larger ones are not cost-effective compared to other energy alternatives. Particularly because of the capital intensity of nuclear power, it seems unattractive for debt-strapped developing nations.[65]

Right now, coal, small hydropower, and cogeneration plants are all much cheaper than nuclear fission.[66] Hydropower is particularly attractive because, although North America and Europe have developed 60 percent and 36 percent of their hydropower potential, respectively, Asia has used just 9 percent, Latin America 8 percent, and Africa only 5 percent. In China, for example, 76,000 small hydropower plants supply almost 10,000 MWe of power in rural areas. And, by the year 2000, according to some experts, cogeneration can account for ten times that amount in some countries.[67]

Biomass is also an inexpensive alternative for rural electrification. The greatest use of biomass residues is found in the relatively treeless plains of Northern India, Bangladesh, and China, where crop residues and dung provide as much as 90 percent of household energy in many villages and a considerable proportion in urban areas.[68] Likewise, in inner Mongolia, 2,000 small wind turbines are used for lighting, running television, electrifying corral fences, and projecting movies.[69]

Compared to such alternatives, nuclear power is also a questionable energy source because it creates fewer jobs and requires more dependence upon foreign companies and governments than almost any other investment a developing nation can make. Also, nuclear energy in a developing nation would be a target for military and terrorist abuse in a politically unstable region. Furthermore it would likely serve only the small minority that uses electricity. It would bypass the majority who rely on fuelwood and charcoal. All these considerations suggest that investment in rural electrification, using small-scale renewables, is a better way for developing nations to go.[70]

8. Conclusion

Apart from the ethical and economical reasons for avoiding the generation of radioactive waste, there might be a number of general philosophical considerations that suggest that the commercial, and not only the military, nuclear path is wrong. On this view, how we deal with the radwaste problem might be a barometer for the collective sanity and morality of our society. A thousand years ago, the world's finest architectural and engineering talents were mobilized to build cathedrals. It is ironic that that comparable talents and even more skills are today dedicated to devising foolproof nuclear garbage dumps.[71]

NOTES

1 P. Z. Grossman and E. S. Cassedy, "Cost Benefit Analysis of Nuclear Waste Disposal," *Science, Technology, and Human Values* 10, No. 4 (Fall 1985): 49; hereafter cited as: Grossman and Cassedy, Cost Benefit Analysis.

2 David Hawkins, Asst. Administrator for Air and Waste Management, *Considerations of Environmental Protection Criteria for Radioactive Waste*, Washington, D.C., US EPA, February 1978, p. 1; hereafter cited as: Hawkins, Considerations.

3 See, for example, J. Raloff, "Nuclear Waste Still Homeless," *Science News* 136, No. 3 (July 15, 1989): 47; R. Monastersky, "More Questions Plague Nuclear Waste Dump," *Science News* 135, No. 25 (June 24, 1989): 389; R. Monastersky, "Opening Delayed for Nuclear Waste Site," *Science News* 134, No. 13 (September 24, 1988): 199; finally, see Zhores Medvedev, *Nuclear Disaster in the Urals*, trans. George Sanders, New York, Norton, 1979.

4 U.S. Energy Research and Development Administration, *Final Environmental Statement: Waste Management Operations, Hanford Reservation, Richland, Washington*, Vol. 1. (ERDA-1538) (Springfield, Virginia: National Technical Information Service, October 1975), p. x-28; hereafter cited as: ERDA-1538.

5 This statement was made by nuclear proponent Gordon Linsley, in his essay, "Radioactive Wastes and Their Disposal," contained in this volume.

6 For information on Maxey Flats, see James Neel, "Low-Level Radioactive Waste Disposal," statement in US Congress before a subcommittee of the Committee on Government Operations, House of Representatives, 94th Congress, Second Session, February 23, March 12, and April 6, 1976, Washington, D.C., US Government Printing Office, 1976, p. 258. See also US Geological Survey, vertical file, "Maxey Flats: Publicity," Louisville, Kentucky, Water Resources Division, US Division of the Interior, n.d.; hereafter cited as: USGS-P. (The Louisville office of the USGS is responsible for monitoring the Maxey Flats radioactive facility.) Finally see A.

Weiss and P. Columbo, *Evaluation of Isotope Migration-Land Burial*, NUREG/ CR-1289 BNL-NUREG-51143, Washington D.C., US Nuclear Regulatory Commission, 1980, p. 5. For data on future US radwaste storage and requirements, see R. Monastersky, "The 10,000-Year Test," *Science News* 133, No. 9 (February 27, 1988): 139-141. See also G. Hart, "Address to the Forum," US Environmental Protection Agency, *Proceedings of a Public Forum on Environmental Protection Criteria for Radioactive Wastes* (ORP/CSD-78-2), Washington, D.C., US Government Printing Office, May 1978, p. 6.

7 US ERDA, "ERDA-1538," Vol. 1, p. x-74. See Chapter 2, notes 13 and 16; Vol. 1, p. II. 1-57. US Atomic Energy Commission, *Comparative Risk-Cost-Benefit Study of Alternative Sources of Electrical Energy*, (WASH-1224), Washington, D.C., US Government Printing Office, December 1974, p. 3-83. See also I. Amato, "Dangerous Dirt: An Eye on DOE," *Science News* 130, No. 14 (October 4, 1986): 221; hereafter cited as: Amato, DOE.

8 This statement was made by nuclear proponent Gordon Linsley in his essay, "Radioactive Wastes and Their Disposal," contained in this volume.

9 J. Raloff and I. Peterson, "Trouble With EPA's Radwaste Rules," *Science News* 132, No. 5 (August 1, 1987): 73.

10 J. P. Murray, J. J. Harrington, and R. Wilson, "Chemical and Nuclear Waste Disposal," *The Cato Journal* 2, no. 2 (Fall 1982): 569; hereafter cited as: Murray, Harrington, and Wilson, Waste Disposal.

11 Sierra Club, *Low-Level Nuclear Waste: Options for Storage*, Buffalo, New York, Sierra Club Radioactive Waste Campaign, 1984, p. 2.

12 J. M. Deutch and the Interagency Review Group on Nuclear Waste Management, *Report to the President* (T1D-2817), Springfield, Virginia, National Technical Information Service, October 1978; hereafter cited as: Deutch, Report.

13 D. MacLean, "Introduction," to D. Bodde and T. Cochran, "Conflicting Views on a Neutrality Criterion for Radioactive Waste Management," University of Maryland, College Park Center for Philosophy and Public Policy, February 23, 1981, p. 3; this publication hereafter will be cited as: Bodde and Cochran, Radioactive Waste Management. See also Deutch, Report, pp. D-11; D-12; D-14, and D-19.

14 W. S. Caldwell, *et al.*, "The 'Extraordinary Nuclear Occurrence' Threshold and Uncompensated Injury Under the Price-Anderson Act," *Rutgers-Camden Law Journal* VI(2) (Fall 1974): 379. See also Kristin Shrader-Frechette, *Nuclear Power and Public Policy*, Boston, D. Reidel, 1983, pp. 10-11; hereafter cited as: Shrader-Frechette, NPPP.

15 Cited by Sheldon Novick in a taped interview with Carl Walske in *The Electric War*, San Francisco, Sierra, 1976, pp. 32-33. See also Shrader-Frechette, NPPP, pp. 8-9.

16 See Shrader-Frechette, NPPP, pp. 75-81.

[17] Christopher Flavin, *Nuclear Power: The Market Test*, Washington, D.C., Worldwatch Institute, December 1983, p. 33; hereafter cited as: Flavin, Nuclear Power.
[18] A. Lovins and L. Lovins, *Energy/War*, San Francisco, Friends of the Earth, 1980), chapters 2-3.
[19] E. Winchester, "Nuclear Wastes," *Sierra* (July-August, 1979); hereafter cited as: Winchester, Nuclear Wastes.
[20] Grossman and Cassedy, Cost Benefit Analysis, p. 48.
[21] Murray, Harrington, and Wilson, Waste Disposal, p. 586.
[22] Hawkins, Considerations, pp. 27-29.
[23] Winchester, Nuclear Wastes. See also R. Schneider and J. Trask, *U.S. Geological Survey Research in Radioactive Waste Disposal - Fiscal Year 1982*, Water Resources Investigations Report 84-4205, Reston, Virginia, US Geological Survey, 1984), pp. 38ff.
[24] Winchester, Nuclear Wastes.
[25] Mark Crawford, "DOE, States Reheat Nuclear Waste Debate," *Science* 230, No. 4722 (11 October 1985): 151.
[26] Flavin, Nuclear Power, p. 31.
[27] Cynthia Polluck, *Decommissioning : Nuclear Power's Missing Link*, Washington, D.C., Worldwatch Institute, April 1986, p. 13; hereafter cited as: Polluck, Decommissioning.
[28] See note 11.
[29] Gordon Linsley (see his article, "Radioactive Wastes and Their Disposal," in this volume) subscribes both to the neutrality criterion and to the "dilution and dispersal" concept. As a result, he is inconsistent.
[30] Polluck, Decommissioning, p. 15.
[31] In their articles in this volume, Linsley and Pasztor both erroneously speak of radioactive-waste "disposal."
[32] R. E. Kasperson, *Equity Issues in Radioactive Waste Management*, Cambridge, Massachusetts, Oelgelschlager, Gunn, and Hain, 1983.
[33] Shrader-Frechette, NPPP, pp. 57-58.
[34] J. J. Berger, *Nuclear Power*, Palo Alto, California, Ramparts, 1976, p. 150; hereafter cited as: Berger, 1976.
[35] For a discussion of this issue, see K. Shrader-Frechette, "Nuclear Arms and Nuclear Power: Philosophical Connections," in M. A. Fox and L. Groarke (eds.), *Nuclear War: Philosophical Perspectives*, New York, Peter Lang Publishers, 1985. See also S. M. Cohen, *Arms and Judgment*, Boulder, Colorado, Westview, 1989, and K. Kipnis and D. Meyers (eds.), *Political Realism and International Morality*, Boulder, Westview, 1987.
[36] Derek Parfit, "Energy Policy and the Further Future," University of Maryland, College Park, Center for Philosophy and Public Policy, February 23, 1981, 19 pages, especially p. 1; H. S. Burness, "Risk: Accounting for the Future,"

Natural Resources Journal 21, No. 4 (October 1981): 723-734; Grossman and Cassedy, Cost Benefit Analysis, p. 51.

37 See P. L. Joskow, "Commercial Impossibility, The Uranium Market, and the Westinghouse Case," *Journal of Legal Studies* 6, No. 1 (January 1977): 119-176.

38 See, for example, Mark Sagoff, "Property Rights and Environmental Law," *Philosophy and Public Policy* 8, No. 2 (Spring 1988): 9-12. See also L. C. Becker, *Property Rights*, Boston, Routledge and Kegan Paul, 1977.

39 P. J. Leahy, US Senator, Vermont, "The Socioeconomic Effects of a Nuclear Waste Storage Site on Rural Areas and Small Communities," statement before the Subcommittee on Rural Development, Committee on Agriculture, Nutrition, and Forestry, US Senate, Washington, D.C., August 26, 1980, Washington, D.C., US Government Printing Office, 1980, p. 2.

40 Flavin, Nuclear Power, p. 31.

41 Berger, 1976, pp. 94-97, 106-112, 144-47. See also J. Gofman and A. Tamplin, *Poisoned Power*, Emmaus, Pennsylvania, Rodale, 1971, pp. 177, 199; J. Primack and F. Von Hippel, "Nuclear Reactor Safety," *The Bulletin of the Atomic Scientists* 30, No. 8 (October 1974): 7; E. Muchnicki, "The Proper Role of the Public in Nuclear Power Plant Licensing Decisions," *Atomic Energy Law Journal* 15, No. 2 (Spring 1973): 45.

42 Flavin, Nuclear Power, p. 41.

43 Flavin, Nuclear Power, esp. pp. 1-33. See also Saunders Miller, *The Economics of Nuclear and Coal Power*, New York, Praeger, 1976, p. 105. Finally, see Shrader-Frechette, NPPP, p. 123.

44 Flavin, Nuclear Power, p. 42.

45 For information about the West Valley facility, see R. E. Kasperson, "The Socioeconomic Effect of a Nuclear Waste Storage Site on Rural Areas and Small Communities," statement at the Hearing before the Subcommittee on Rural Development of the Committee on Agriculture, Nutrition, and Forestry, US Senate, 96th Congress, Second Session, August 26, 1980, Washington, D.C., US Government Printing Office, 1980, pp. 61-62; hereafter cited as: Kasperson, 1980. For substantiation of the claim that poorly educated, financially strapped people/communities are those that consent to locating a risky facility nearby, see note 48.

46 See Amato, DOE; see also Dr. Alice Stewart's study cited in Winchester, Nuclear Wastes.

47 Flavin, Nuclear Power, esp. pp. 40-42.

48 K. Shrader-Frechette, *Risk Analysis and Scientific Method*, Boston, D. Reidel, 1985, pp. 107-112; hereafter cited as: Shrader-Frechette, RASM. K. Shrader-Frechette, *Science Policy, Ethics, and Economic Methodology*, Boston, D. Reidel, 1985, p. 137; hereafter cited as: Shrader-Frechette, SP.

49 Shrader-Frechette, SP, p. 214.

50 Shrader-Frechette, SP, p. 215.

[51] Shrader-Frechette, SP, pp. 214 ff. See R. B. Stewart, "Pyramids of Sacrifice? Problems of Federalism in Mandating State Implementation of National Environmental Policy," in F. A. Strom, (ed.), *Land Use and Environment Law Review-1978*, New York, Clark Boardman, 1978, pp. 162-163. See also R. B. Stewart, "Paradoxes of Liberty, Integrity, and Fraternity: the Collective Nature of Environmental Quality and Judicial Review of Administrative Action," *Environmental Law* 7, No. 3 (Spring 1977): 472. Finally, see M. Markovich, "The Relationship between Equality and Local Autonomy" in W. Feinberg, (ed.), *Equality and Social Policy*, Urbana, Illinois, University of Illinois Press, 1978, p.96.

[52] See Shrader-Frechette, RASM, Section 3.3.2 of Chapter Two. See C. Zracket, "Opening Remarks," in Mitre Corporation, *Symposium/ Workshop...Risk Assessment and Governmental Decision Making*, McLean, Virginia, The Mitre Corporation, 1979, p. 3. (The Mitre publication will hereafter be cited as: Mitre, Symposium/Workshop. See C. Starr, "Benefit-Cost Studies in Socio-technical Systems," in Committee on Public Engineering Policy (ed.), *Perspectives on Benefit-Risk Decision Making*, Washington, D.C., National Academy of Engineering, 1972; and D. Okrent and C. Whipple, *Approach to Societal Risk Acceptance Criteria and Risk Management*, PB-271 264, Washington D.C., US Department of Commerce, 1977; A. P. Hull, "Discussion," in Mitre, Symposium/Workshop, pp. 171-172; and N. Rescher, *Risk: A Philosophical Introduction*, Washington D.C., University Press of America, 1983, pp. 35-40.

[53] C. Starr and C. Whipple, "Risks of Risk Decisions," *Science* 208 (June 6, 1980): 1114-1119. See also C. Starr, *Current Issues in Energy*, New York, Pergamon, 1979, p.15.

[54] Shrader-Frechette, SP, p. 134. See also Kasperson, 1980, pp. 62-63.

[55] J. C. Lashof, *et al.*, Health and Life Sciences Division of the US Office of Technology Assessment (OTA), *Assessment of Technologies for Determining Cancer Risks from the Environment*, Washington D.C., OTA, 1981, pp. 3, 6ff.

[56] Quoted by Derek Parfit, *Reasons and Persons*, Oxford, Clarendon Press, 1984, p. 511.

[57] E. Peele, "Innovative Process and Inventive Solutions," p. 6, unpublished manuscript, 1986, available from Peele at the Energy Division, Oak Ridge National Laboratory, Oak Ridge, Tennessee 37831; hereafter cited as Peele, Innovative Process. See also E. Peele, "Mitigating Community Impacts of Energy Development," *Nuclear Technology* 44 (June 1979): 132-140, esp. pp. 133-36.

[58] W. A. O'Connor, "Incentives for the Construction of Low-Level Nuclear Waste Facilities," in *Low Level Waste*, Final Report of the National Governors Association Task Force on Low-Level Radioactive Waste Disposal, Washington D.C., 1980. E. Peele, "Siting Strategies for an Age of Distrust," unpublished essay presented at the US Department of Energy, Argonne

National Laboratory, June 30, 1987, and available from Peele at the address listed in note 57; finally, see Peele, Innovative Process.

59 Bodde and Cochran, Radioactive Waste Management, p. 7.

60 Polluck, Decommissioning, p. 27. Flavin, Nuclear Power, pp. 45-47. For substantiation of this point, see the del Callar essay in this volume.

61 Flavin, Nuclear Power, p. 31.

62 Polluck, Decommissioning, p. 27.

63 Flavin, Nuclear Power, p. 15.

64 Flavin, Nuclear Power, p. 26.

65 Flavin, Nuclear Power, p. 52.

66 Flavin, Nuclear Power, pp. 54-59.

67 But much of this includes reliance on natural gas and some dependence on coal. See Christopher Flavin and Alan Durning, *Building on Success: The Age of Energy Efficiency*, Washington, D.C., Worldwatch Institute, March 1988, p. 35.

68 Jessica Tuchman Mathews, *World Resources 1986*, New York, Basic Books, 1986, p. 111.

69 Kosta Tsipis, "Nuclear Power and Energy Needs of the Third World Economies," *Church and Society*, Report and Background Papers, Meeting of the Working Group, World Council of Churches, Glion, Switzerland, September, 1987, pp. 227, 229.

70 Flavin, Nuclear Power, pp. 54-55.

71 Winchester, Nuclear Wastes.

Nuclear Energy and the Transformation of Creation: A Theological Comment

Peter Sedgwick
University of Hull
Hull, England

Any discussion of nuclear power, including that of the churches, must now reckon with the debate in Britain in the last three years on the privatization of the electricity industry. This debate concerns the alleged compatibility among market forces, energy production, conservation, and safety. Therefore, any Christian reflection on the nuclear issue (a "theology of nuclear power") cannot be concerned simply with the technology of the industry, or even the details of the environmental crisis, such as the precise reasons for global warming. Instead, the moral theologian is forced inevitably to consider such issues as, "what place do market forces have in energy supply and conservation?" "Is energy production concerned with its reliability, safety and frugality, or is it also intertwined with the competitiveness of a national economy?" "How would shareholders act differently from a public-sector body in handling questions of decommissioning nuclear stations?"

Clearly, the debate has become political. Its arguments are different from those of the 1960s onwards, which were pro- or anti-nuclear. The new debate concerns what we mean by the "polis," the public forum in which issues of energy are resolved. In this chapter, I cannot face all these questions. Instead I wish first to look at some of the issues which have come up in the British privatization debate in the last two years. Second, there is the difficult issue of whether public debate is actually possible. Some would argue that because of capitalism, or the mass media, or the sheer diversity of this world in the late twentieth century, there is no meaningful ethical dialogue any longer. I wish strongly to resist this pessimism. In the third section, I raise specifically theological issues. This section is not related especially to nuclear power. Nevertheless, I wish to make a fundamental point. It is, I believe, central to the Christian tradition: we are charged with transforming the environment. This does not mean that we destroy, damage, or inflict wanton harm, but nevertheless we do change it. Deep at the heart of the nuclear-power debate is the issue of whether humanity can have the right to shape the created earth for the well-being of human life. Therefore, while the third section is not specifically to do with the nuclear issue, I believe that the basic question of material transformation determines all else. Whether nuclear power is the most ecologically sensitive way of handling material transformation in Africa, or elsewhere, is a very contentious point. In the final two sections I raise questions about planning and sustainability.

A longer chapter would consider the nature of modern capitalism, the question of the market, the position of countries in Africa (where the 1989 WCC consultation took place), and both the activities of multinationals in the nuclear-engineering industry, alongside the efforts of technologists in countries such as Argentina, to provide cheap energy supplies through nuclear power. Inevitably this chapter is rather general, and remains at a somewhat abstract level. My plea in mitigation is that this chapter discusses the ethical and religious framework within which the debate takes place. Clearly, having built a house, the need is to decide how people are to live inside it. All this chapter does is point to the basic architecture of moral discourse, within which theologians, environmentalists, technologists and public-policy experts can conduct their (sometimes heated) debates. And even this is perhaps to claim too much, for some would say that this is a Western, Christian architecture of moral reasoning about human creativity. But it is at least one strand in the inheritance of the World Council of Churches.

1. The Market and Nuclear Energy

The consensus among experts is that carbon-dioxide emissions may double in the next thirty years.[1] But to maintain the present degree of pollution in the atmosphere, the carbon-dioxide emissions need to be halved. One strategy, much used in the US, is to employ prices and taxation of electricity-generating companies in a way that encourages energy conservation. Economists in Britain and North America are now searching for regulations that will achieve the greatest conservation for the least distortion of the market. There is no consensus as to whether nuclear power can play a part in reducing carbon-dioxide emissions. It is not simply an environmental/safety argument; there is also the market factor, because increasing nuclear-power generation will require both vast investment and a high concentration of scientific talent to run reactors successfully.

The British Central Electricity Generating Board (CEGB) has, for a while, conceded that nuclear power could not compete economically with a privatized electricity industry. Even before the withdrawal of nuclear power from the flotation of electricity generation, the CEGB admitted, at the Hinkley-Point inquiry into a proposed PWR, that the economic argument had moved against nuclear power. It is certainly possible to argue for the fact that economic growth can bring certain goods: it provided, for instance, the introduction of widespread central heating in Britain. A rise in living standards, because of such heating systems, contributes to lowering child death rates as much as does sophisticated medical technology. While endless economic growth is clearly to be questioned, there is a moral case to be made for the benefits brought to Western Europe by the economic growth of the last four decades. The environmental opposition to economic growth in the 1950s-1970s failed to deal with this point.

Nevertheless, the supply of plentiful energy, which fuelled the economic growth of the post-war era, cannot be met by nuclear power. There are major difficulties in combining safety, the market, and energy supplies. Admittedly, western technology has standardized the light-water reactor, but it was created by state intervention for supplying weapons material and for submarine propulsion. In order to be adapted to electricity generation, reactor designers added many safety systems which proved unreliable at Three Mile Island and Chernobyl. Even if they were fully reliable, there is no reason why market demand could create an "inherently safe" reactor. An author writing in *New Scientist* states:

> The High Temperature Gas-Cooled Reactor now waits in the wings. It is unlikely to emerge into the centre stage of energy production until demand for power once again outstrips supply. Even then, it may languish, unless government sources invest additional billions of dollars to demonstrate the practical usefulness of new generations of gas-cooled reactors.[2]

Yet the British CEGB is cutting back on nuclear research all the time, especially on fast-breeder reactors. Again, there is a conflict among the demands of safety, those of the market, and those of nuclear research.

> No one denies that fast-reactors offer long-term security of power supply. But they are expensive. The industry, faced with privatization, believes it cannot afford to fund the research. The Government apparently believes that any research the industry does not want to support, cannot be worth supporting....The nuclear episode is a particularly egregious example of how the long-term planning necessary to modern high-technology projects is incommensurable with the priorities of politicians.[3]

There is no simple technological answer to the twin problems of energy need and environmental responsibility. Energy is inherently costly, in financial, social, and environmental terms. Environmental auditing is no longer merely concerned with anti-pollution legislation.[4] It also addresses environmental impact and the relationship of the community to energy generation. Therefore, the nuclear issue must be resolved within the context of the failure of the free market to support the growth of nuclear power in the last four decades. In Britain, at least, the free market is now rapidly withdrawing support from research in nuclear energy. The *New Scientist* article, already cited, points to similar difficulties for research in West Germany and the United States. Should such research go on? It will not contribute to plentiful, cheap energy supplies, at least not in the short run. So the 1990s will be dominated by the cost of energy: environmental, social, and financial.

Who bears the cost of such energy burdens in society? Who decides how a civil society is to be formed, with certain areas (like nuclear generation of electricity) set aside from the play of the free market? Scientists can no longer assume that they are regarded with unquestioning awe and acceptance by society. Nor can politicians simply make decisions which will be implemented without vociferous protests. So the theologian and philosopher must ask whether there can be moral agreement in society at all. We move to the second part of the paper.

2. Is Moral Agreement Possible?

Some have argued that the answer is clearly "no." Partly this is because moral agreement now has to be sought internationally. However, international agreement involves the discussion of Western/Eastern, Northern/Southern perspectives, which may be impossible to reconcile. So the first argument against possible agreement is that of relativism, the different cultural norms found worldwide. A second is the failure to agree on what moral terms mean. It is significant that, in the first part of this paper, I introduced a non-moral use of the term 'good'. Economic growth brought higher living standards and lower child death rates in the last forty years. But what is "good" about that? The non-moral sense is obvious, but morally this leads one into a discussion of human well-being. Are lower abortion rates "good"? It depends on one's perspective. So the second reason for the alleged failure of moral discourse is the breakdown in moral agreement, indeed, a breakdown even in what it means to be moral. The third reason is the perceptible tendency to leave such questions to the experts, or the state, and get on with one's own private life. Let me examine each of these in turn:

2.1 *Cultural Relativism*

It is highly ironic that, in debates on the philosophy of religion, the Christian theologian is on the defensive to prove the unchanging nature of the concept of humanity, while in this (energy) debate the concept of a human perspective which is universal has been unchallenged. Challenges to the universality perspective come either from the perspective of time (across the centuries) or space (what is true in one continent is not true necessarily in another). This argument is common to much liberal discussion of the doctrine of salvation, for instance; human beings thought and were different twenty centuries ago, or see things in a different perspective today in different cultures (the famous duck/rabbit paradigm perspective).

If one moves to the question of the importance of tradition in many cultures, even today (as asserted by many anthropologists), there is the issue of whether the basic concepts, norms, and presuppositions in different cultures can be reconciled (e.g. the relationship of the nature of personality to the community in which a person is set; understandings of relationships, dignity, and worth). This issue of cultural relativism is highlighted in a recent observation:

> Traditions are often one of the major impediments to change in any society and in the rural areas of the developing countries traditions are particularly

> strong. Whilst some traditions are clearly anachronistic, others are reflections of an underlying rationality that has evolved over many years to cope with the problems of survival in an often harsh environment. The fact that they have persisted is, in a sense, proof that they have been (on the whole) successful.
>
> Traditions are often deeply ingrained in the way of life of rural populations, playing an important role in maintaining cultural integrity and community cohesion.[5]

While the author (just quoted) uses cultural symbols in an over-functionalist way (claiming that "somebody is of value if it helps society work," rather than asking "what is the meaning-world of this society?" as contemporary philosophers of religion might do), the general point is correct: cultural relativism is often an impediment to progress.

My contention is not that we must accept the constraints of cultural relativism and say that each culture (in time and space) thinks differently about human destiny. Rather, my contention is that agreement among cultures must be fought for in discussion, not taken for granted. There is perhaps more that separates persons, one from another, in individual and corporate cultural identities, than unites us.

2.2 *The Breakdown of Moral Discourse*

Here I clearly draw on the work of philosopher Alisdair MacIntyre,[6] as well as that of his past colleague and theologian Stanley Hauerwas.[7] The argument is put succinctly in a parable:

> There was once a highly developed civilization with an impressive grasp of science, both in its understanding of pure research and applied technology. An explosion destroyed this civilization and all who lived in it. Centuries later the area was re-colonized by a community which found thousands of pieces of paper containing scientific data. The community learnt and recited such pieces of paper, recognizing the vast wisdom of a past civilization which far outstripped their own. But it made no sense (except to themselves in the act of recitation) and nothing worked technically. For they had no idea of how the pieces of paper related, or what the underlying scientific theory and methodology was on any of the subjects described in these ancient sheets. (paraphrased)[8]

The claim made by Hauerwas is that we have lost an overall moral tradition, a sense of what moral values mean, and how these relate to underlying assumptions about what it means to be human. Cultural pluralism need not entail social breakdown, yet it has occurred anyway. Concepts like honor, purity, and sin are almost literally meaningless now in the West. MacIntyre argues this way, as does Hauerwas, because he believes that there are no longer any institutions which can carry from generation to generation the meanings and values which once were embodied in social roles and occupations. Furthermore, he argues that western political discourse degenerates rapidly into slogans, not because of the dehumanizing technology of the mass media, which reduces all thought to thirty-second "bites," but because the sense of moral discourse is not sufficiently understood in society to warrant further exploration. It is not that there are no philosophers or theologians who could provide an understanding of moral discourse; there are plenty. But most of them no longer interact with the majority of those who influence our societies across the West. The philosophers are following the religious experts into a sanitized ghetto. Two points are crucial:

- When moral discourse breaks up, people continue using the language, thinking they can relate to each other, "if only the other person sees things from my perspective." They do not realize that the fundamental, overall perspective has dropped out of sight. Moral polarization results.
- There is a regress to two different positions in moral debate. First, technical and statistical data are far more cited in the debate. We cannot decide what we mean by a healthy society or individual, so we quote deaths diverging from the SMR (standardized mortality rate) in an area. We debate the figures, but do not relate the figures back to the ethically substantive issue. (This is very common in political debates, which can turn into lists of figures). Second, since we are uncertain what it means to be moral, we only know that we have the freedom to be moral, and that freedom becomes the most important thing for us. Any social constraint, self-accepted or imposed from outside, becomes anathema. MacIntyre points to the liberal western intellectuals, condemned to be free by their own choices, in everything from relationships to beliefs, but with no ability to see that freedom is "for something." Here MacIntyre argues that religious and cultural traditions (often the same tradition) may offer a "way out," in a world "after virtue," though the danger of a collapse into religious fundamentalism is equally real and utterly disastrous. MacIntyre's influential work is marked by a deep pessimism about the possibility of sustained moral discourse, though other philosophers would still assert the existence of a common moral sense. Whether this is anything more than a psychological point, or a pragmatic/empiricist position, is unclear.

2.3 Social Legitimation

A very different tradition, though still concerned with moral value, is that of the neo-Marxist Frankfurt school, and its successors.[9] Here the emphasis is on social development. The growth of the modern technological elite has reduced moral questions to issues of technology. How do we handle issues of energy, economic growth, and so on? People are excluded from the debate, because of their failure both to participate in the bureaucracy and to understand the technology. Increasingly, political debate is presented as having the nature of a "technical fix." The result is the demise of substantive ethical questions about "what sort of society do we wish to create?". Voter participation drops, and politics ceases to be a debate for the citizens of the polis/city, which is all of us, and instead becomes a minority interest, like gardening or art. At the same time, technology is promoted by modern industry into a mass commercialization of modern culture. People earn more and more (for a majority in the West, though not elsewhere, and not for a minority in the West). Yet the culture in which they participate is less and less imbued with any moral values. This point is repeatedly made in Habermas' social analysis. In a more detached way, John Habgood's own writings also chart the difficulty of a culture's handing on moral values from one generation to another.[10] Habermas points to the rise of single-issue groups in social life, groups that do not engage in the totality of political life. Between mass commercialization and technology, between pre-packed slogans and an impenetrable bureaucracy, the ordinary citizen turns away from politics into a de-politicized world. The final point is reached where there appears no alternative even conceivable to the mass-consumption, western culture for its inhabitants. "Wasn't it always like this? Aren't those people over there really sad they aren't part of it? Of course they are. We must figure out a way to bring them in too."

3. A Theological Response to Moral Disagreement

The moral theologian does not have to accept this pessimism. Moral reasoning is possible, even if there is no consensus and probably no overall moral tradition. There is enormous concern, especially among young people, that their exploration of creativity, personal relations, and well-being should be extended to include both other people and the creation. Such concern often operates in a very immediate way and looks for rapid solutions, yet there seems to be the possibility of a fragile alliance between the few who have a grasp of ethical traditions and the many who express a deeply grounded concern. Equally one could point to the concern about violence, child abuse, and sexual stereotyping, or the intense anguish about human rights, political well-being, and pollution that has erupted in the Soviet Union and Eastern Europe in the last eighteen months.[11] The free market, which now operates throughout western society, will probably result in a more mobile, faster changing culture, where values are easily relativized or consumerized. Despite all of this, there remains the possibility of raising environmental and religious concerns. The relevance of these concerns to the energy debate is illustrated by the following questions.

- What do we mean by a just society, and will that discussion be the same in each culture? Can divergent views be reconciled? Is there a "Third World" at all?
- Is the energy crisis not a moral issue, but a prudential one? A person lighting a can of petrol to cook her dinner is to be dissuaded on prudential grounds. It will blow her (and us) to bits. Sharing the dinner with a hungry neighbor is a moral question, involving ethical considerations, such as: "why should I share it; why don't they work for it too?" Which way are we debating the

energy debate? I suspect it is the former. If it is the latter, then what ethical criteria are we using, and how do we handle the moral disagreements which are inherent to a moral debate? So back to Macintyre (note 7).

- How is the energy debate related to the emergence (in the West) of a post-industrial, social-market, liberal-democratic society? This is the area of my chief interest as a social ethicist and moral theologian. How do energy issues get raised in the media, in political debate, in terms of power and participation? In Britain the last few years have been dominated both by environmental concerns about energy, and by the privatization of first oil (BP; British National Oil Corporation), then gas, now electricity generation (including nuclear power) and, if this government stays in power, finally coal. How does the market handle energy issues, and how does a market-led energy policy relate to public debate and participation? Back to Habermas (note 9).
- What is the inter-generational ethical issue, especially in terms of a shared set of moral assumptions and values that we may or may not hand on to our children along with the technological problems? (See Habgood, note 10, and Shrader-Frechette's paper, this volume.)

A just society will be one in which the participation of its members is encouraged, and different possibilities of human flourishing are made possible. This does not rule out many competing understandings of human relationships, some of which may appear retrograde to western eyes, such as the growth of Islamic fundamentalism. Here there must be room for vigorous moral debate, but the western view of technology and individualism is not the only way forward. Yet, in a just society, there will have to be curbs on the power of the market. It is by developing particular campaigns and pressure groups, both of which seek to influence public opinion while still recognizing their own limitations, that the subtle interplay of market forces and moral demands will be worked out.

4. Creation, Vocation and Technology

Some would argue that "there is no specifically 'Christian' way...of deciding for or against the nuclear generation of energy."[12] Instead, they claim that the Christian Gospel is effective when it imprints upon Christian men and women a certain character which enables them to participate in the discussion, and carry out their task of responsibility and vocation. The Judeo-Christian tradition shows the ambiguities of human nature, thus contextualizing particular claims about nuclear energy.

Others would point to the development of a Christian doctrine of vocation, where God calls her people to be co-partners in the task of creating a community of moral value and purpose. (I take it that metaphorical language about the nature of God allows the use of either masculine or feminine provisions for God, so long as the understanding of what the metaphor is conveying is kept in view.

I am not advocating an indiscriminate gender typology.) Previous decades before this one have seen the gradual emergence of a doctrine of vocation for industrial society where the nature of work was given meaning by the use of language about human creativity and cooperation. Partly this language failed to reflect the sheer boringness of manual factory work or the coercive nature implicit in production-line assembly work; partly it laid unnecessary stress on the masculine attributes of strength, or used the common scientific metaphors of "victory over creation," "conquering scientific problems," or other similar terms. For whatever reason, a Christian doctrine of vocation at work has become more problematic in recent years. Alongside this, of course, has come the rise in mass unemployment in some parts of the West, the awareness of under-development in other parts of the globe where paid work is not part of the expectation of every adolescent, and the move to a post-industrial society, which perhaps implies more leisure time.

All this leaves the major question unanswered, however. What meaning does our daily life under God actually have, whether it is a life of work, leisure, or unemployment? Unless we can say why we wish to work, we cannot say why we need industrialization, and hence why we need energy supplies at all.

At this point perhaps some biblical exegesis may be of help. In recent decades there has been a high degree of interest in the wisdom tradition which is found in the Old and New Testament. Such a tradition expresses a belief in the fundamental given-ness of the world and its moral order. Especially in Job, Proverbs, some of the Psalms and the Apocrypha, this tradition is clearly delineated. It is not an alternative to the prophetic tradition, because the prophets also sought to recall Israel to the demands of the law and a moral way of life. But if prophecy was closely related to finding revelation in worship of Yahweh as Lord of Creation (the royal Psalms where Yahweh is spoken of as king in the Jerusalem temple), then the vision of wisdom found revelation in the ordering of creation itself. Ultimately, however, the two strands come together. Israel keeps agreement with God (covenant) out of gratitude for the sheer grace of the gift of the promised land, flowing with milk and honey (Deuteronomy 8:8 and 11:14 articulates the ecological sense of blessing), while "the wise" praise God for the goodness of the created order.

This created order is made by a dynamic God. Within God's own being, there is a rich dynamism, so that wisdom always rejoices before God. This is only a personification of the divine attribute of creativity, but it sets out the overflowing abundance within God, which means wisdom rejoices in "the inhabited world," delights in "the sons of men." Such a dynamic God has a purpose for the creation, set out in Psalm 8 and Genesis 1, where that purpose is given to Adam. But lest we move too easily to human co-partnership with God, note that the wisdom tradition speaks of the glory of the dynamic creation, where the wild animals range free, "scorning the tumult of the city, deaf to the shouts of the driver" (Job 39:7). God speaks to Job out of the whirlwind, and God overwhelms Job. This is a highly dynamic view of creation.[13]

Creation is the embodiment of moral values, spoken of by the wisdom tradition as light and not darkness. Such light is continually re-established by God and will not be denied; even if "I take the wings of the morning, and dwell in the uttermost parts of the sea, even there thy hand shall lead me" (Psalm 139:8-10). Creation of God is for a purpose: God delights in the creation, builds up the moral order in and through creation, and calls humanity to care for this order. Even with evil and human wickedness, "as long as the earth endures, seedtime and harvest... will never cease" (Genesis 8:22).

What, then, is the charge to humanity? "We go forth to our work, and to our labor until the evening," but, in a nice touch, the morning rush-hour meets the rush-home of the lions getting back to their dens before the sun rises. All this is part of God's wisdom, where the Spirit renews the face of the earth. This is the theme, of course, of the WCC Assembly at Canberra. Psalm 104 asks God to rejoice in the creation as the psalmist both rejoices in the Lord and prays for the end of wickedness.

The charge to humanity is to co-operate in the dynamic activity of God, as God orders in a dynamic way the creation. God is the ultimate energy event, a pure activity of love and rejoicing in creativity. How do we co-operate? As the natural creation is affirmed in and for itself, humanity must care for creation in and for itself. Second, creation embodies moral values in potentiality. Basic to the wisdom tradition is that moral value (enshrined in the Torah or Law) and the given-ness of creation are both orderings of reality. Both are dynamic. Moral value depends on the lordship of humanity over the creation, while ever respectful of it. As the fruits of the earth and human physical activity interact together, the biological and physical are transformed (without negating the prior state) into the moral and spiritual. Matter and human labor make a house. A house shelters human beings and animals. A house becomes a home for a family. The Law is read in the family. The Jewish family praises God and goes back to care for the earth and to transform it. It is a spiritual eco-system of wisdom building a house (using feminine attributes) with seven pillars, in which people "walk in the way of insight" (Proverbs 9:8). Wisdom was there "before the beginning of the earth" (Proverbs 8:22) as the hills emerged and, at the end, as the wise person goes to give advice to the rulers, and the family tends its crops (Proverbs 11:14/27-23). It is all one system, moral and natural.

The dynamic activity of God calls, then, for the preservation of creation and its transformation from the material into the moral and spiritual. Nor is this view simply rational. Again and again, the wise person considered the emotions as fundamentally just, as crucial as the mind or the will. The creation of deep emotional bonds enables the bearing from one generation to another of moral obligations and spiritual values. Finally, this history is open-ended: there is the promise of freedom for all creation, the end of suffering and pain, in a narrative of promise and hope. In Jesus of Nazareth this is spoken of as the coming of the Kingdom of God, which is a symbolic expression of the rule of God over all creation and its response in penitence and thanksgiving.

Protestant theologians at the Reformation, especially in the Lutheran tradition, were insistent on the vital importance of re-capturing the sense that every person, as a Christian or not, had a vocation under God to order creation. Earthy as ever, Luther spoke of washing nappies/diapers as part of the vocation of a Christian family. What is more problematic is the clear division of that vocation into work, the State, and the family as different aspects of one vocation for humanity. It is problematic because technology and the complexity of life have run all these areas together. Work is transformed by automation, the State by computerized information; the family by *in vitro* fertilization and modern medicine (such as pharmacology). There are now no limits, apparently, to the promise or threat which technology holds.

Technology clearly has gone too far in damaging creation. However, my original question was on the meaning of human activity. Does technology help or hinder the creation of moral values, emotional bonds, and social relationships? Here is the Christian contribution to the debate. Vocation is about the way that any activity, paid or unpaid, can create a dynamic moral and spiritual order, where the particularities of a person are valued, but seen also as a contribution to the totality of a community's tradition. Technology can be part of that vocation.

So Section Two, which described the moral crisis of the late twentieth-century industrialized society, is not answered by Section Three. That would be too easy a solution. But for Christians and Jews, and for many associated with that tradition who seek their own spiritual destiny, the resources of the Judeo-Christian tradition will still be the source and inspiration of their answer to the moral crisis. How technology helps the transformation of creation and its preservation, rather than its annihilation, will be a difficult question to answer. Nevertheless, there is an answer to the question about why we need energy. We need energy to meet our vocation in creating and sustaining a dynamic moral order. Part of that vocation includes the material transformation of the environment surrounding us. Yet, if that destroys the environment as well, the vocation is self-defeating. We cannot move back to a pre-industrial order, but there are two challenges to a theological reflection on industry. One is how to control its environmental effects. The other is to find in industry (and in the post-industrial world we now inhabit) the bearer of meaning and value in its interaction with matter. Both are fundamental to the wisdom tradition. Both are still relevant today.

5. A Planned Society and Christian Hope

An affluent society is often a fragmented one. Personal provision is made for one's own family, irrespective of social consequences. Yet ultimately no one can escape being shaped by, and shaping, our society. The nature and form of a society is shaped by its energy use, its transformation of matter, and the types of raw material that it consumes.

If we want assured and increasing energy supplies, if we believe that our current rate of technical expansion can continue as at present on a worldwide

> basis without leading to exhaustion and collapse, then we have to pay the price in terms of technical and social complexity. And the character of our society will be formed by the extremely sophisticated technology received to secure its primary resources. Energy supplies shape communities.[14]

The question therefore arises as to how decisions on energy supplies ought to be made. How far can public participation be made possible? There is no possibility of participation if the truth is not known. In recent years, sin and wrongdoing have been described theologically both individually and in corporate or institutional terms. Thus a system which hides the truth might be called "sinful." Again, we return to the wisdom tradition's juxtaposition of light and darkness, as metaphors for the moral order. Equally, the Fourth Gospel dwells on the relation of truth (alethia) to light and freedom (8:31-2). Truth here is seen as an unveiling of what is hidden. Yet much of the energy debate is marked by secrecy over several decades, especially on the civil use of nuclear energy.

Even if they obtain knowledge, communities may be powerless. David Bell's definition of power is a linguistic one: "If you do x, I will y." Y represents a sanction in terms of threat or promise. Influence statements have a different syntax: "If you do x, you will do y." Here a series of predictions are made about future consequences of a proposed course of action. They are sanction-free. No community ever simply operates in terms of influence-statements, but it is characteristic of many Third-World societies that all negotiations about energy with the developed world are backed by implicit sanctions of threats or promises. They are often not in a position to refuse the promise or avert the threat. The relationship is therefore a power relationship. This is also true of many poor communities in the developed world, like the former mining communities in England.

Christian hope sees the potential of science as part of a God-given creativity, a creativity that can be traced back ultimately to the wisdom tradition. This speaks of participation by humanity in God's wisdom. But such potential can be dangerous. John Paul II said at Hiroshima, "We know that this potential is not a neutral one: it can be used either for man's progress or for his degradation." The glorification of technology in the earlier part of this century had effects on communal existence, effects that tended to be a glorification of speed, of the individual and of centralized decisions. Today we seek, in a more difficult moral context, to keep our societies in the West open and participatory, while allowing the underdeveloped world more freedom from the constraints of unequal negotiations on energy needs. It is not a comforting prospect, although the situation is complex and not fully determined. Can there be a moral restraint on the use of science? This leads us into the final section of the chapter. It deals with the notion of a sustainable society.

6. A Sustainable Society

Human creativity works well only within a given order. The world is given an order by God, but within this order, there is a deep randomness that contributes to its dynamic. Scientific thought, for example, recognizes the Uncertainty Principle of Bohr and Heisenberg. Furthermore, the consequences of any decision exceed the personal life-span of any individual. As technology imposes yet more awesome questions of responsibility and provides more options for human life, the complexity of a technological society becomes ever more difficult to handle. This is so, even if we also assert that energy supplies determine that community as well.

There is no point in time, given the nature of the universe created by God, that one can say that human nature has ever inaugurated the future by the technology which it creates under God. Yet the logic of that technology may lead us to believe that we can in fact control the future, and that it is a moral imperative to do so. How is this so?

If technology is inaugurated, there may be an assumption that, when mistakes occur in the future, then they can be overcome by further planning. Alterations in the program have further implications, and a set of responses, therefore, are built in an infinite regress. This may be the case, but there are many studies of past environmental damage which was irrevocable, and yet where its perpetrators assumed that the technology at fault could be modified. (I am not thinking of the damage caused by, for instance, coal fires, where in the last century there was no way of modifying the technology.)

The philosopher Stephen Clark, in an unpublished paper, distinguished between health and safety.[15] In personal health, we are not considered healthy if we need a whole chemical industry to keep us alive, although chemical products may be quite safe. By analogy, if the environment can only be kept going, after technological damage to the environment, by managing it with further technology and our continuing presence and concern, then it may be a safe environment, but not a healthy one. The argument is not that technological damage renders human beings unhealthy, but that it causes the environment itself not to be healthy. Of course, there may be a trade-off between technological benefits, such as short-term answers to energy needs in the underdeveloped world, and the health of the environment. Nuclear waste renders an environment unhealthy: it needs permanent monitoring. Yet nuclear energy may yield benefits. (I refer to the future nuclear energy programme: we clearly have to deal with existing waste.) For generations, humanity depended on, and interacted with, the environment, as the wisdom tradition shows clearly. It is symptomatic of this century, not only that the environment has become dependent on humanity, but also that a moral case has to be made out against this. Again I come back, as so often in this paper, to the difficulty both of constructing a moral case and of persuading others. Perhaps only the imminent threat of mass pollution will move people. This is the analogy of setting light to the can of petrol.

Stephen Clark also refers to Job and Psalm 114. He writes:

> We want to see a healthy nature, but a genuinely healthy nature is one that can get on without our constant care. Even if the technosphere were to

> contain a regulation patch of "wilderness," a few tolerated beasts, it would not meet our requirement for a world that will continue without our assistance, that does not (in the last resort) need our hand on the tiller. It is the sheer indifference to us and to our projects displayed by the rest of the natural world that attracts us (see Job 38:41; Psalm 114), for this indifference is our evidence that things are, after all, going smoothly, that Gaia (the terrestrial biosphere) is not yet a terminal case.[16]

At this point, the argument could well move into a consideration of intergenerational ethics, as in Shrader-Frechette's paper in this volume. I recognize the importance of this dimension, but prefer to stay with the relationship of planning to sustainability. The argument concerns restraint in the development of technology. Christianity is deeply concerned about restraints imposed on oneself, out of a response to the Creator, and about the ordering of life. ("For everything there is a season, and a time for every matter under heaven" (Ecclesiastes 3:1). "Forbearing one another" (Colossians 3:13).) Scientific research also can be subject to such restraint. The scientist then involves the policymaker in the application of research to social needs. Again, there is a subtle interplay between the two positions. The scientist has an optimistic motivation, justifying the research and showing its social utility. The policymaker is cautious, and perhaps even pessimistic. The possible dangers must be weighed, and all the consequences uncovered before a decision is made, with as much public participation as possible. The policy adopted must be flexible, and not simply mandated, regardless of what new dangers may be revealed. This is not an anti-scientific attitude. Energy policy is a long-term, continuous process, with prudence and restraint as an inevitable part of that process. Hence questions like global warming and nuclear waste all come into play.

Responsible planning may advocate the new generation of small nuclear reactors in a world of limited energy supplies, or it may reject nuclear power because it has historically been allied to a high-growth, centralized society which is no longer appropriate. There is a tension here. The contribution of theology is to point to the need for openness, participation, and responsibility. Openness includes, in Karl Rahner's words, not just freedom of information, but openness to the future. Christianity, he wrote, is the guardian of the openness of the future.[17] Technological planning can become over-confident, so that the future is thought to be under our control. Rahner wrote:

> If a point were ever arrived at which the future were really understood as something which could be comprehensively prognosticated - at that point the future would, basically speaking, be eliminated in favour of a continually abiding present.[18]

Over-planning allows the substitution of an unhealthy, safe environment for a healthy one, and assumes that technology can be added on to increase safety. (This was especially true of earlier nuclear-reactor designs.) Under-planning neglects the challenge of energy needs, or calls for a regression to a pre-industrial society. True planning hopes for a future in which humanity recognizes the challenge given by God to care for the earth and to transform it, meeting the needs of a hungry and developing world. It also hopes for a future given by God. We sustain the fragility and sensitivity of our world.

7. Conclusion

I have resisted the notion that the creation is the source and origin of the sacred. Rather I have sought to place the dimension of the sacred in God, with dynamic order at the heart of God. Such a God in the Judeo-Christian, and especially wisdom, tradition delights in the creation and in an equally dynamic moral order. Both the moral order and the ecological balance do not seem in very good shape at the moment. The challenge to Christian theologians is both to affirm their own traditions and to enter into worldwide dialogue with other traditions. Equally, moral value must be re-affirmed where there seem to be few living traditions, but only a complex diversity. Out of this moral complexity, society has to fashion the means to change public attitudes about energy and the environment. I have concentrated on two areas in the debate. Both concern planning. One is the question of public participation, and the other is that of long-range forecasting and the design of technology. Theology cannot answer these problems, but it can raise questions which may point to some of the future directions where the energy debate could succeed. Perhaps the final point is to allude once again to the complexity of the issue, as the wise always worked, "weighing and studying...with great care" (Ecclesiastes 12:9). This is no academic retreat into learning for its own sake. The wise moved cautiously, since they knew that God watched over the creation, judging every deed, whether good or evil (Ecclesiastes 12:14). It is God's creation which we are given to care for, and we answer for our actions.

We answer for our actions in the market-place of the world, where the power of the free market often feels overwhelming. "A rich man's wealth is his strong city, and like a high wall protecting him." (Proverbs 18:11). (Interestingly, the variant reading on this verse is "protecting his imagination.") The task of the theologian and ethicist is not to let the market rule our imagination, but to form new ways of balancing the costs of energy, the demands of society, and the fragility of creation. There is no easy technological fix to this balance, but the possibility of the "inherently safe" reactor must be weighed against the increasingly ruthless search for profits in the energy industry and the ever-growing awareness of environmental damage. The definition of a good person, by the wise person, was that there should be "an inheritance for his children's children." (Proverbs 13:22) In the last decade of this century, we must recognize the moral imperative of long-term caring for creation. This is part of the wisdom that God grants to his children. (Job 28:28)

Notes

1 See the chapter by Janos Pasztor, "What Role Can Nuclear Power Play in Mitigating Global Warming?" in this volume; see, especially, section one of Pasztor's chapter.

2 Dan Charles, "Nuclear Safety: Some Like It Hot," *The New Scientist* 124, No. 1689 (November 4, 1989): 58-62.

3 Tom Wilkie, "An Energy Policy in Chaos," *The Independent* (October 31, 1988): 17.

4 "How Green is My Company?" *The Economist* 314, No. 7645 (March 10, 1990): 128.

5 David O. Hall, "Biomass for Energy in the Developing Countries," *Anticipation* 29 (November 1982): 34.

6 Alisdair MacIntyre, *After Virtue*, London, Duckworth, 1981; Alisdair MacIntyre, *Whose Justice? Which Rationality?* London, Duckworth, 1988.

7 See the discussion of MacIntyre in Stanley Hauerwas, *The Peaceable Kingdom*, London, SCM, 1984; hereafter cited as: Hauerwas, PK.

8 Hauerwas, PK, pp. 4-5.

9 J. Habermas, *Legitimation Crisis*, trans. T. McCarthy, London, Heinemann, 1976.

10 John Habgood, *Confessions of a Conservative Liberal*, London, SPCK, 1988.

11 M. Edwards, "Siberia, in from the Cold," *National Geographic* 177, No. 3 (March 1990): 31, 34-35.

12 See, for example, Jordon Dunston, "Theological amethod in the Deterrence Debate," in G. Goodwin (ed.), *Ethics and Nuclear Deterrence*, London, Proom Helm, 1982.

13 Cf. Dan Hanly and David Ford, *Jubilate: Theology in Praise*, London, Darton, Longman and Todd, 1983.

14 John Habgood, A *Working Faith*, London, Darton, Longman and Todd, 1980, p.104.

15 Stephen Clark, "The Household of Earth and Heaven," unpublished paper, Society for the Study of Theology, Annual Conference, 1982; hereafter cited as: Clark, Household.

16 Clark, Household.

17 Karl Rahner, *Theological Investigations*, Volume 12, London, Darton, Longman and Todd, 1974, p. 200; hereafter cited as: Rahner, TI.

18 Rahner, TI, p. 200.

WCC Church and Society Working Committee Report

1. Introduction

The Working Committee endorses the ethical guidelines proposed by the 1989 Kinshasa Consultation, noting that they have long been part of the WCC agenda. It commends them to all those involved in policymaking on nuclear energy.

In spelling out their significance, the committee interprets respect for "the sustainability of creation" as entailing avoidance of any irreversible damage to the environment. Such a policy must maintain adequate energy options for future generations on a worldwide basis, and it must not leave future generations with problems of waste and pollution which carry unacceptable risks.

The Committee interprets a just energy policy as entailing a fair distribution of energy resources which takes account of the different energy needs and wants of different countries.

At the Committee's request one member wrote a scenario of what such a just and sustainable policy might mean in practice. This is appended to the report as the expression of a personal view which the Committee, while not endorsing, believes is of great interest.

2. Participation in Decisionmaking On Nuclear Energy[1]

Many people have strong feelings about the political and cultural context in which discussions of participation in decisionmaking take place. Indeed, it is arguable that the shapers of energy policy neither understand, nor choose, energy options in a political or ideological vacuum. As policymakers, they are people in positions of power, influenced by others in power. All too often, they determine energy policy without taking into account the needs and wants of those whose lives will be affected by their decisions. In most instances, the latter have not been a part of the decisionmaking process, nor have they been provided with the information necessary to play such a role. Indeed, particularly with respect to nuclear energy, there has often been an air of secrecy and a political climate of centralized control that obstructs shared decisionmaking. The secrecy and centralization are often justified, rightly or wrongly, by concerns for security.

Moreover, it is arguable that, whether they realize it or not, policymakers, like everybody else, are guided by worldviews, by ways of perceiving reality, that serve as a background for their policy deliberations. These worldviews give the policymaker a sense of what is, and what is not, real; of what is, and what is not, knowledge; of what is, and what is not, the nature and appropriate aim of human life; of what is, and what is not, the value of the natural world apart from its usefulness to humans. These worldviews also influence the ends toward which technology aims.

The influence of worldviews raises certain questions for the nuclear-energy debate. What has been the worldview in terms of which, heretofore, the debate has been articulated by policy shapers? Has this worldview been influenced by the mechanistic paradigms of classical science, in which the physical world has been viewed almost exclusively as a resource for human use, lacking value and purposes of its own? Has it been one in which human beings have been understood primarily as individualistic consumers of energy, rather than as persons-in-community? Has it been one in which economic development has been understood primarily in terms of increased production and consumption, rather than as the promotion of human community in a just and sustainable context? Has it been one in which the appropriate knowledge required for participating in the nuclear debate has been deemed scientific knowledge? Have other kinds of knowledge, for example, the experience of disadvantaged and traditional peoples, concerning nature and community, been neglected? Has this worldview been one in which, wittingly or unwittingly, Western values are deemed superior to others? If such a worldview in fact has been a backdrop for much energy discussion to date, can and should an alternative worldview, perhaps shaped by dialogue between Western and non-Western voices serve as a backdrop for the foreseeable future? In any consideration of nuclear options, these questions merit sustained attention.

3. Nuclear Energy and Technical Issues[2]

Both experts and the public regard reactor safety and long-term waste storage as the greatest area of concern regarding commercial nuclear energy. (This excludes, by definition, the issue of diverting reactor fuel for weapons manufacture). The Committee notes efforts to improve the safety features of reactors. Essentially, as summarized in the paper by J. Pasztor (this volume) the aims are to minimize further the risk of core meltdown and strengthen the physical containment of reactors. Given the lessons from Three Mile Island and Chernobyl (and many other less serious reactor incidents) which indicated the prime importance of human errors, much attention is being given to reducing those mechanical and electronic operational components dependent on human control. Another line of approach is to re-consider core design in order to incorporate features of certain reactors with long records of relatively safe operations (e.g., former gas-cooled and Candu models). Existing fast-breeder reactors, however, pose especially serious potential problems associated with controlling fission in their cores and with their sodium coolant systems. On safety grounds alone, their future development should not be supported.

3.1 *Questions of Safety in Nuclear Energy*

There are different safety criteria for an "acceptable" reactor. An optimal set of criteria, according to J. Pasztor (this volume) includes "...a core which will neither melt down nor catch on fire upon any conceivable operator action or sabotage, including abandonment of the site. Furthermore, such a reactor should be able to hold its radioactive inventory following attack with explosives, e.g. by a terrorist group or an earthquake." This probably is a desirable, though idealistic, prescription. No reactor could be designed which would withstand extreme assaults upon its integrity.

Other reactor problems include safety of routine radioactive emissions and escapes of material during the normal operations of reactors. Here it is important to review results of recent epidemiological investigations into the possibility of an association between low levels of radiation and harmful effects on humans. Of particular concern are the recent studies on the clustering of young persons suffering leukemia who lived in the vicinity of some British reactors (up to ten times the number of cases by comparison to the frequency of leukemia in the general population). The scientific conclusion (Valerie Beral, in *Epidemiology and Radiation Protection*, Nuclear Energy Agency, Paris, 1988, p. 128) is that "these (cases of leukemia) cannot be dismissed as chance findings, although the reasons for the excess are far from clear. The doses of radiation received by the children are estimated to be too small by several orders of magnitude to explain the excess. This paradox needs to be resolved." This situation requires resolving, especially because of other incidents of unexplained associations, notably those related by Joyce Hollyday (this volume) that persist among people living near Three Mile Island (and also include pathological occurrences among plants and animals).

Safety standards for exposure to ionizing radiations are currently under review, and it is expected that the International Committee on Radiological Protection will produce a revised set of standards in late 1990 or early 1991. These are likely to require that the exposure limits be made more strict, probably by a factor of two or three.

Much more effort is needed to adequately train, and update the expertise of, nuclear workers in the safety aspects of their work. In some countries, most of the nuclear workforce exposed to the highest risks is drawn from poorly educated and otherwise disadvantaged sectors of the population. Further, it is questionable whether such workers actually make a decision under free, informed consent to work in an industry with special hazards and with less restrictive radiation standards than those for the general public.

There is an inequity in the siting of reactors by countries on or near the borders of another country (e.g., by the French near Geneva, by the Swedes near Denmark and by the Czechs near Austria). Such sitings create an unfair risk to those who do not receive any energy benefit from the other country's reactors or, as in the case of Austria, may have decreed by democratic means their opposition to such means of electricity generation. The matter would best be resolved by choosing sites according to the principle that the groups of persons who benefit should approximate to the group of those who face the greatest potential hazards (i.e., who live nearest to a reactor).

We condemn the dumping of radioactive and toxic chemical wastes in developing countries. Action is urgently needed for binding international agreements to prevent hazardous-waste dumping in developing countries (incorporating IAEA radioactive waste storage guidelines together with those of WHO for toxic chemical waste). The IAEA should have its responsibilities expanded to include powers for monitoring the transfer of nuclear waste across national boundaries and also for assisting member countries to train personnel capable of operating radioactivity-detection equipment in ports and airports. The IAEA should publish annual reports on transboundary movements of waste and on its management.

Other nuclear hazards concern the risks of exposure (especially to radon gas) during uranium mining and milling. A major problem is *ensuring* adherence to safety regulations. Also, independent (non-nuclear industrial) monitoring of all radiation hazards should be guaranteed and funded by governments.

3.2 Prospects and Problems For Managing Nuclear Wastes

Although some progress has been made in recent years with burial techniques for low- or intermediate-levels of radioactive waste, there are as yet no permanent high-level waste facilities in operation. The issue generates great concern because of public exposure to the risk of environmental contamination. This occurs through escape of dangerous material through faulty design, human errors, terrorist activities or sabotage, earthquake, and breakdown of the controlling institutional authorities. On the latter point there are differing views. One view is that final engineered solutions will be found that require no major arrangements for ongoing surveillance. The other view is that, without ongoing institutional control, the long-term prospect of hazardous human intervention cannot be excluded.

There is no alternative other than to proceed with developing techniques for permanent storage of high-level waste currently contained in temporary storage. Prospects for continuing to produce increasing quantities of high-level wastes are, however, subject to dispute. Essentially, there are two conflicting arguments. Both camps acknowledge the issue as a serious one, subject to great public concern. One argument is that technological advances and the efforts being made to achieve them indicate that final disposal solutions would be a reality for the near future. Accordingly, there are no grounds for suspending the operations of reactors because of inability to solve problems with high-level wastes.

The other point of view is based on the principle that we have no right to force future generations to live in an environment which we knowingly have put at risk through imposing on it a potentially serious hazard to future human well-being. Especially in countries where there exist viable alternatives to nuclear power (as in the USA, where the proportion of electricity generated by solar, hydro, wind, and other soft technologies exceeds that generated by nuclear reactors), there is an extremely strong ethical case for progressively excluding reactors from the energy portfolio and expanding alternative energy resources. By such means future generations will avoid having the long-term hazard of high-level waste forced upon them. On this view, nuclear wastes created by electricity generation are an unacceptable burden for future generations, both because their containment cannot be guaranteed, and because their presence denies to future generations their in-principle rights to redress and to free, informed consent. Accordingly, this line of argument supports the phasing out of nuclear power on the basis that ethical and economic grounds will outweigh future benefits in countries that have a capability for substituting alternative electricity sources. It should be noted, however, that there are definite environmental hazards associated with alternative energy sources.

3.3 *Questions of Design for Future Reactors*

Even if design improvements could guarantee greater safety, the long lead-time for bringing on-line new reactors means that it would be over a decade before they were available for commercial use. Moreover, the whole process would be extremely costly.

3.4 *Evaluation of Full Costs of Nuclear Energy*

Estimating nuclear costs is a very difficult issue. As regards economic costs, there are well-known problems associated with hidden military expenditures during the initial development phase of civilian nuclear power industries in nuclear-weapons nations and the existence of many direct and indirect governmental subsidies. Of greater concern, and essential for any rational cost-benefit consideration, are the non-economic costs--relating to environmental impacts, health, and life-style (the latter including restrictive legislation designed to protect nuclear activities). Benefits of nuclear energy include equitable availability and economically valuable usage of electricity. Costs include all activities of the nuclear fuel cycle, from uranium mining to waste disposal, and from reactor construction to decommissioning, including the externalities.

The methodology is difficult, especially for estimating non-economic costs. Methodological progress has been made in rich, industrialized societies, on the principle of calculating opportunity costs for different energy sources in order to draw comparisons and assign priorities. Nevertheless, the general applicability of such methods, which include putting a monetary value on a human life, are questionable. Difficult interactions with other societal arrangements, for instance, with political and social organization, legal structure, and the cultural acceptability of sanctions and rewards built into the cost-benefit structure of nuclear systems, also impede calculating full nuclear costs.

It is doubtful whether the means exist for a universally applicable evaluation of the full costs of nuclear energy. Efforts should continue towards the development of such methods as an important step towards a rational and informed planning process.

3.5 *Access to Information Concerning These Subjects*

There are gross problems in the area of access to information, due to breaches of the principle that the public has the right to information about risks. Many recent instances reveal the historical nuclear-industry practices of systematic propagation of disinformation and deception. This malpractice extends from silence about well founded risk estimates (e.g., that there is a one in four risk of core meltdown during the average 30-year life-span of existing commercial US reactors) to denial of the basic right to have access to information about personal harm (e.g., to the medical records of a radiation-exposed person). We regret the official explanations for denial of information, for instance, claims that the information would threaten "national stability" (i.e., economic interests) or "security" (with reference to a civilian electricity-generating enterprise). Government authorities are often reluctant to discharge statutory responsibilities, not only where freedom-of-information legislation exists, but also in failing to support epidemiological investigations, despite widespread allegations of biological damage and public anxiety.

We reject the frequently used claim by authorities that release of information would cause mass panic. There should be no exclusion whatsoever of information about the use of nuclear power to generate electricity. Furthermore, much more needs to be done to inform and educate the public about all aspects of nuclear-power industries. We appreciate, however, that this policy may cause difficulties in countries where there are cultural or other barriers to the dissemination of information.

3.6 *Making Ethical Decisions in Situations of Uncertainty*

We acknowledge that there are also great difficulties because of scientific uncertainty. Where no final technological solution has been identified, as in the case of high-level nuclear waste disposal, there is a profound degree of uncertainty with estimating any consequences, especially when public safety regarding some nuclear wastes must be guaranteed for millions of years. Where risk estimations are derived on the basis of differing ethical principles, conclusions may differ considerably, for instance, when using a utilitarian (for the majority) principle or an egalitarian (for everybody) one. We believe that, in the utilitarian case, where some minority has been left at a disadvantage (e.g., displaced to make way for a nuclear facility or subject to "greater than average health risk") considerable attempts should be made to provide as fair and equitable solution as possible. This would involve seeking free, informed consent from the public, negotiating conditions for acceptable risk with potential victims, and compensating them in full for whatever risks are imposed on them. If this is not obtained, then the best available alternatives should be put in place. Such may include monitoring systems to detect adverse impacts on health or well-being or a system of adequate and appropriate compensation. We believe that ethical decisions should be incorporated in forward planning and not adopted after legal actions force recognition of some adverse consequences. Facilitation of ethical decisionmaking should follow construction of a proper forum, conducted under judicial or quasi-judicial arrangements, and in which all parties in the issue are properly represented to argue their points of view. Non-governmental organizations with a good track record in such negotiating processes may prove valuable allies, especially in cultures where the people may be intimidated by the process or unfamiliar with participating in it. There are some authoritarian societies that will neither agree to any process involving public participation, nor attempt to achieve fair and equitable solutions for those disadvantaged by decisions made by decree. All that can be said is that authoritarian decisions face creating a legacy of human suffering and despair that will persist for decades--as has occurred in the nuclear arena for the indigenous people of the north Marshall Islands, New Mexico, Arizona, and Central Australia.

4. Nuclear Energy and Theology[3]

In biblical accounts of creation, there is a diversity of pictures describing the relationship between God, humans, and the rest of creation. Genesis (1:26-28) says that humanity is created in God's image to have "dominion" and "subdue." While some interpretations have justified attitudes of human domination over nature, the Hebrew understanding of dominion "in God's image" meant humanity's presence in creation to carry out God's rule, not for human ends, but for God's ends. In the biblical story of the flood and the re-establishment of the world, these words are repeated but without the injunction to subdue (9:1-7). Moreover, there is repeated reference to the covenant between God and all the earth (9:9-13). Several biblical references assign to humans the role of "tending" the earth, as one would tend a garden. In Psalms 104 - also a creation story - the whole of creation and its interdependence, including humanity, is sustained by God's Spirit or breath (Ps. 104:30). Many passages, such as this, underscore the human capacity for interfering with the intended goodness and harmony of God's creation. Further, biblical pictures from the wisdom tradition emphasize the awe and mystery of creation, as well as the human search to discover God's presence and purposes within the created order. The understanding of redemption, biblically, refers not just to redemption of human beings, but to the redemption of all Creation by God through Christ.

4.1 *Attitudes to Creation*

This variety of pictures leads to various interpretations of the relationship between humankind and creation, including, at times, a focus on dominion and mastery. An understanding of "stewardship" (tending the garden) has also been emphasized. Some see this as implying a separating and managerial relationship between humans and the creation. Others see this more inclusively. Another perspective stresses creation's integrity and the interdependence of humans and the created world.

God's Spirit is present within creation. We express this as the "sanctity" or "mystery" of creation. All creation is thus worthy of our celebration, praise, and wonder. There is an interrelatedness and interdependence among all the parts of creation. They form an integral whole. Humankind has a unique responsibility towards creation. Humanity participates in creation and shapes it through its choices which can nurture or destroy. Acknowledgment of a responsibility entails the need for re-examination and repentance in attitudes and actions. Inherent in the concept of responsibility is the necessity for participation.

A prevalent attitude of humanity has been to modify and dominate nature. This entails standing apart from nature, using it simply for selfish ends without regard for the consequences. As a result, human actions often violate the integrity of creation and today endanger its very survival.

In acknowledging the damage done to creation as a result of this attitude, we recognize the need for rediscovering the insights of theology regarding the relationship between humanity and God's creation. Humanity's search for knowledge, which is part of human responsibility under God, should be directed by a sense of the blessedness of creation and an understanding of its integrity. This can lead us to new and creative solutions.

4.2 *Technological Choices Reflect Attitudes to Creation*

Human attitudes of domination over nature have shaped our choices of technology. Equally, the choices we make about energy technologies and their uses will shape the societies in which we live. Specific applications of various technological systems may promote the common good or destroy it. Paradoxically, technology can create isolation or enable communication, increase freedom or narrow choices, destroy or nurture life. We can place too much hope in technology, thinking that every technological problem can be solved with another "technological fix."

Theological perspectives can influence energy choices. Human attitudes of domination, often justified by certain theological interpretations, have sometimes resulted in a choice of energy systems with little or no reflection on their environmental or social impacts. A theology of stewardship suggests that energy choices take into account resource depletion, as well as environmental, social and economic impacts. Theological perspectives which stress the relationship between human beings and the rest of creation, underscoring the integrity of creation and our responsibility towards it, would indicate making energy choices in light of their inherent effect on creation. These energy choices will promote a participatory and environmentally sustainable society.

4.3 *Wholesome Attitudes to Creation and Technology*

Different cultures have different perceptions of reality and will see their energy needs and resources in different ways. Nevertheless, we suggest some general ethical principles which may be applicable to any culture. These are:

- the responsibility to future generations and to promote the sustainability of creation;

- justice as enabling human survival and fulfillment; and
- participation of people in energy choices which directly affect their lives.

We recognize that all harmful energy producers cannot be stopped instantly, but recommend that countries move as rapidly as possible towards implementation of all three of the above principles. We urge that the priority of future energy research be given to those technologies that best express these three ethical principles.

Notes

1 One sub-group at the Kinshasa consultation discussed the issue of participation in decision-making on nuclear energy. Because the sub-group was unable to to reach a consensus, the Working Committee has attempted in this section to summarize the main questions discussed by this group.

2 Approved by the Kinshasa consultation, this third section of the Working Committee Report is a slightly edited version of the Report of the Working Group on Critical Issues, Technical Questions, and Debate. Members of this Kinshasa group were Bena-Silu, Charles Kerr, Gordon Linsley, Freda Rajotte, Kristin Shrader-Frechette, and Liek Wilardjo.

3 Approved by the Kinshasa consultation, this fourth section of the Working Committee Report is a slightly edited version of the Report of the Working Group on Theology. Members of this Kinshasa group were Carlos Araoz, Wesley Granberg-Michaelson, Ann Heidenreich, Joyce Hollyday, and Peter Sedgwick.

Appendices to WCC Church and Society (CS) Working Committee Report

Participants, 1989 Consultations and Meetings of the World Council of Churches

Members, CS Working Committee of WCC

The Most Rev. Dr. John HABGOOD, *England*
Moderator of the Committee

Prof. T. K. OOMMEN, *India*
Vice-Moderator of the Committee

Bishop Galinski FEOFAN, *USSR*
Executive of the Committee

Dr. Jesse MUGAMBI, *Kenya*
Executive of the Committee

Dr. Mady LAEYENDECKER-THUNG, *Netherlands*
Executive of the Committee

Dr. BENA-SILU, *Zaire*

Prof. Charles BIRCH, *Australia*

Prof. Sergio BROFFERIO, *Italy*

Rev. Susan EAGLE, *Canada*

Mr. Masaki ENDO, *The Philippines*

Prof. Dr. Heino FALCKE, *Democratic Republic of Germany*

Prof. Dr. Karol GABRIS, *USSR*

Dr. Antony HOUBANCHEV, *Bulgaria*

Ms. Darlene KEJU-JOHNSON, *Marshall Islands*

Prof. Dr. Theodor LEUENBERGER, *Switzerland*

Dr. Jay MCDANIEL, *USA*

Dr. Rolando MENDOZA, *Costa Rica*

Ms. Pirkko PENTTILAE, *Finland*

Ms. Lilia O. RAMOS, *The Philippines*

Dr. Ulrich RATSCH, *Federal Republic of Germany*

Mr. Smith Ngulh Za THAWNG, *Burma*

Prof. Kosta TSIPIS, *USA*

1989 Zaire Consultation Participants

Dr. Carlos ARAOZ, *Argentina*
Fuel-Cycle Division, Comision Nacional de Energia Atomica

Dr. BENA-SILU, *Zaire*
Eglise de Jesus Christ sur la Terre par le Prophete Simon Kimbangu

Dr. Achilles del CALLAR, *The Philippines*
Adamson University

Ms. Dorothy GOLDEN-ROSENBERG, *Canada*
A Directory of Women in Canada Specializing in Global Issues

Ms. Ann HEIDENREICH, *Kenya*
Environmental Liason Centre International

Ms. Joyce HOLLYDAY, *United States of America*
Sojourners Magazine

Professor Charles KERR, *Australia*
University of Sydney, Department of Public Health

Ms. Corrine KUMAR D'SOUZA, *India*
CIEDS

Mr. G. S. LINSLEY, *Austria*
International Atomic Energy Agency

Professor MALU-WA-KALENGA, *Zaire*
Commissioner for Zaire Nuclear Energy Programme

Dr. B. C. E. NWOSU, *Nigeria*
Research and Technology Ltd.

Dr. Ulrich RATSCH, *Federal Republic of Germany*
Forschungsstatte der ev. Studiengemeinschaft (FEST)

The Rev. Dr. Peter SEDGWICK, *United Kingdom*
Church of England, University of Hull, Department of Theology

Dr. Kristin S. SHRADER-FRECHETTE, *United States of America*
University of South Florida, Department of Philosophy

Ms. Karin STAHL, *Federal Republic of Germany*
Forschungsstatte der ev. Studiengemeinschaft (FEST)

Dr. Liek WILARDJO, *Indonesia*
Satya Wacana Christian University

Ing. YOKO, *Zaire*
Director General NEC

WCC Consultation/Sub-Unit Staff

Wesley Granberg-Michaelson, Director

Freda Rajotte, Programme Secretary

Peony Wong, Administrative Assistant

Personal Views

Comments on the Future Role of Nuclear Power in the Sustainable Development of a Just World

Kosta Tsipis

Since the last time the WCC considered the issue of nuclear power, there has been clearer manifestation of several factors which affect future energy policy. The environmental crisis, long ago predicted, has now begun to manifest itself in tangible ways. The depletion of the ozone layer over the Antarctic - much more severe than anticipated - has now *shown* that human activities *can* and *do* affect the environment on a global scale. We had already witnessed environmental distress on a continental scale in the form of acidification of lakes and streams and the destruction of forests due to acid rain caused by the burning of coal to produce energy.

There is some evidence and there are strong predictive analytical models which indicate that the concentration of carbon dioxide in the atmosphere is increasing at such a rate that it can alter the climatic conditions of the Earth in the span of a few years.

There is no experimental evidence of this phenomenon (the greenhouse effect) as of yet but if it begins manifesting itself in an observable fashion, it will be too late to attempt its reversal. We must then *anticipate* its appearance by taking steps to avoid it. Even if the scientific predictions about the imminence of the onset of the greenhouse effect were inaccurate, prudence dictates that we should act in a "worse-case-analysis" fashion and adopt energy policies to counter the possibility of the phenomenon's emergence. The greenhouse effect will have such a profound impact on the planet's life that we can simply not chance its onset by overlooking the probability of its happening.

There is tangible potential conflict between environmental preservation and employment patterns in many parts of the world: In order to save the forests of the northeastern United States of America and Canada, we would have to shut down the coal-burning industrial activities in the Ohio Valley. But that would idle ten million workers. Similar situations can be duplicated in western Europe, the central Soviet Union, and many areas of China.

Overall the least socio-economically disruptive approach to environmental preservation is not to reduce the production of energy but to produce it in environmentally sound ways. Broader socio-economic arguments point in the same direction: Energy is essential to the improvement of the standard of living of all but the most developed regions of the Earth. The emergence of the micro-electronics industry offers the possibility of vigorous economic growth to many developing and under-developed countries without the need for massive energy requirements as was the case with the more traditional industrialization patterns of the past. Relative to past experiences, developing nations now need less energy per unit increase of their GDP. But additional energy to what they produce now, they shall need all the same.

We have then to consider the problem of energy production in the context of sustainability and justice: Environmental concerns must be satisfied, and at the same time affordable energy must be made available to developed nations (which, after all, have caused much of the environmental distress by their energy-production methods and requirements). In considering, then, the future role of nuclear energy on our Earth, we must keep these manifest facts in mind.

Production of usable energy by means of nuclear fission is not inherently more dangerous then coal-burning. However, the technology currently used in nuclear-energy production, the management of the nuclear industry in many nations, and the neglect - until the present - of serious efforts to deal with the highly radioactive wastes of nuclear reactors, have made nuclear-energy generation unnecessarily risky, controversial, and unusually expensive. As a result, nuclear energy has turned from a promising source of cheap energy to an undesirable alternative to energy production by the burning of fossil fuels. The current design of nuclear reactors has the undesirable trait of depending for its safe management on highly trained and disciplined human operators. The record bears this out: The US Navy has operated about 20,000 reactor years without an accident because of its highly trained crews. The US nuclear-energy industry experiences an accident of some sort about every few hundred reactor years and a major one every 10,000 reactor years. In addition, shoddy operations of civilian power-generation reactors have often resulted in undetermined amounts of contamination of the environment with as-yet-uncertain effects to the surrounding human habitat and its occupants. These are not intrinsic properties of nuclear energy. A new generation of reactors has now been designed which is both more suitable for energy generation for civilian purposes and is safer in that it does not depend on human-operator intervention for its safe functioning. The new reactor shuts itself down in case of malfunction, and its cooling system provides for very rapid removal of heat from its core, reducing drastically the possibility of a meltdown at theThree Mile Island or Chernobyl. In combination with fundamentally different industrial practices, and the emerging possibility of safe, but costly, disposal of nuclear wastes, the new type of reactors could contribute to the production of energy in a sustainable and just world.

Nuclear energy safely produced will be expensive, suitable for areas with high concentration of demand for electricity and will require large initial capital investments. As such, nuclear energy can be the energy of choice for developed, highly industrialized areas of the world. At the other end of the energy use spectrum, sparsely populated rural areas in middle latitudes require small-scale local sources of electrical energy which can be optimally provided by photovoltaic solar generators and wind energy. Photovoltaic energy-production technology is now almost at the threshold of being competitive with fossil-fuel energy generation. A vigorous research effort can bring this non-polluting energy source to practical fruition.

Neither of these two energy production technologies, however, are suitable in the case of developing areas of the world. They require amounts of energy and levels of electrical power not easily achievable by photovoltaic technologies, but developing nations do not have the resources - economic as well as manpower - to use the new, safer (but expensive) nuclear technologies. In the name of justice, these areas should be allowed by international agencies charged with environmental regulations to produce energy by methods that burden the environment. While the CO_2 part of this burden is unavoidable, technologies which remove the other pollutants from coal-burning energy generation are available and can be used in these areas of the Earth.

A prudent energy policy, then, firmly delineated by the requirements of sustainability and justice *would* consist of nuclear-energy production in areas of high-density electricity demand in affluent nations within the framework of existing international agreements, fossil-fuel energy generation in developing areas which must not be burdened by the extra costs of nuclear technology, and extensive use of photovoltaic and wind-energy generation for areas of low-density electricity demand. A mixture of energy-producing technologies which will vary in space *and* time appears to be our best option in the quest for a safe and just human community.

None of the technologies mentioned here are quite ready for easy and extensive use. I therefore suggest that we recommend:

- Intensive research and development efforts to complete the development testing and evaluation of the new type of nuclear reactors. At least ten years are needed for this. Similar research and development efforts are needed for the promising deep geothermal energy-extraction technology, and for methods for the clean burning of fossil fuels.
- Continued research and development of photovoltaic technologies, and their rapid dissemination to prospective uses.
- Continued research and development to lower the cost of safe storage of nuclear wastes for periods of about 1,000 years.
- Continued public education and new public-policy initiatives, mainly in the areas of the world with high per-capita consumption of energy, in favor of energy conservation and more efficient end-use of energy.

Statement

G. Linsley and C. Araoz

Some members of the meeting did not agree with all aspects of statements in the report of the consultation. Specifically they disagreed with inferences that nuclear energy presents exceptional risks to humans and their environment.